THE WHISPERING POND

ERVIN LASZLO has written over 50 books, many of which have been translated into a variety of languages, and some 300 papers and articles. He is founder and director of the General Evolution Research Group, founder and President of the Club of Budapest, Past-President of the International Society for System Sciences, and Advisor to the Director General of Unesco.

Essential Society (1963)

Beyond Scepticism and Realism (1966)

System, Structure, and Experience (1969)

Introduction to Systems Philosophy (1972)

The Systems View of the World (1972, revised ed. 1996)

A Strategy for the Future (1974)

The Inner Limits of Mankind (1978)

Systems Science and World Order (1983)

Evolution (1987, revised ed. 1996)

The Age of Bifurcation (1991)

The Creative Cosmos (1993)

Vision 2020 (1994)

The Choice (1994)

The Interconnected Universe (1995)

ERVIN LASZLO

THE WHISPERING POND

*A Personal Guide to the
Emerging Vision of Science*

ELEMENT

Boston, Massachusetts • Shaftesbury, Dorset
Melbourne, Victoria

First published in hardback in 1996
by Element Books, Inc.

This revised edition published in the UK in 1999 by
Element Books Limited
Shaftesbury, Dorset SP7 8BP

Published in the USA in 1999 by
Element Books, Inc.
160 North Washington Street
Boston, MA 02114

Published in Australia in 1999 by
Element Books and distributed by
Penguin Australia Limited
487 Maroondah Highway, Ringwood,
Victoria 3134

Cover design by Mark Slader
Printed and bound in the USA
by Courier Westford

British Library Cataloguing in Publication
data available.

Library of Congress Cataloging in Publication
data available.

ISBN 1 86204 362 0

Endorsements for

THE WHISPERING POND

"A rich overview of the evolution of consciousness and the challenges facing science as it seeks to understand the mind. A wonderful synthesis, full of wisdom and hope."

Peter Russell, author of *The Global Brain Awakens*
and *The White Hole in Time*

"Ervin Laszlo is that complete human being in whom knowledge, faith and social consciousness form a Trinity. To be a passenger on his voyage of revelation, discovery and morality is an opportunity none of us can forgo today nor could ever resist if we tried."

Yehudi Menuhin

"A magnificent book . . . rich and inspiring . . . a beautiful exemplar of how science can add new meaning and richness to our lives."

Raine Eisler, author of *The Chalice and the Blade: Our History, Our Future* and
Sacred Pleasure: Sex, Myth, and the Politics of the Body

"Ervin Laszlo, once again, has put his finger squarely on the critical issues. The tools now in human hands which will shape destiny in the next century, are too powerful to be used by anyone without understanding the meaning of *The Whispering Pond.*"

Edgar Mitchell, Apollo 14 astronaut and author of
Psychic Exploration and *The Way of the Explorer*

"With astounding incisiveness and clarity Laszlo propounds a breathtaking vision."

Karan Singh, President of the International Centre for Science, Culture
and Consciousness, Jammu and New Delhi

"Ervin Laszlo's theory of the psi-field is one of the boldest scientific theories of recent years . . . I maintain that this is of the greatest importance for contemporary science as a whole."

Ignazio Masulli, Professor at the University of Bologna

"This work . . . is one of the first—and to my knowledge the best—to give us a global vision of the evolution of knowledge in this century."

Jean Staune, Secretary-General of the Université Interdisciplinaire de Paris

". . . a most compelling intellectual journey."

David Loye, The Institute of Futures Forecasting, Carmel

"Ervin Laszlo has made important strides in the understanding of the phenomenon of planetary consciousness which is a new evolutionary factor of first importance and will save our own human future and the future life of the planet."

Robert Muller, Chancellor of University for Peace, and former UN Assistant Secretary-General

". . . an inspiring overview, takes the reader to the threshold of the new world being begotten by science at the dawn of the third millennium and shows how it points to a deeper logic subtending all living matter which we ignore at our peril."

Federico Mayor, Director-General of Unesco

"An exciting, optimistic and accurate account of current accomplishments and trends in science. Written by one of the foremost futurists, this book is a must for an informal audience that wants a preview of where we are headed."

Karl H. Pribram, Stanford University

"[Laszlo's] grand unified theory is one of the most important intellectual achievements of our time."

Stanislav Grof, author of *Beyond the Brain* and *The Holotropic Mind*

"Laszlo has accurately put his finger on the single most critical political issue of the next few decades, namely: what picture of reality shall dominate the planet and who shall control it? No one can afford to be unaware of the profound issues dealt with here."

Willis W. Harman, President of the Institute of the Noetic Sciences

"Laszlo may well be the key integrator of the worlds beyond Newton, time and space. *Tour de force.* We can thank Ervin Laszlo for putting our world back together."

Donald Keys, President of Planetary Citizens

"Ervin Laszlo has contributed many of the seminal ideas that will help humankind make the transition to a new century . . . [his] vision of a self-creating cosmos is one of unity, balance, and co-operation. It could impact every aspect of society and culture, serving as a welcome antidote to the torn fabric of our era."

Stanley Krippner, author of *Personal Mythology* and *Spiritual Dimensions of Healing*

"Awesome! A view of the universe that can sustain the human soul in the exalted realms for which we are destined."

Thomas Berry, author of *The Dream of the Earth* and *The Universe Story*

"In these pages, follow the intellectual adventure of one of the great minds of the late twentieth century toward a unified vision of the cosmos."

Alan Combs, author of *Synchronicity* and *The Radiance of Being*

"This book could be written only by someone possessing great courage and wisdom. Laszlo has both. This is a fitting book for the beginning of a new millennium, because our future—if we are to have one—must rest on this or a similar vision."

Larry Dossey M.D., author of *Healing Words* and *Rediscovering the Soul*

Contents

FOREWORD

I

DAVID LOYE

Co-Director, Center for Partnership Studies,
Carmel, California

O NE OF THE hallmarks of this century has been the increasing breakdown of the scientific "verities" of the modern era. Within the past decade or so we have been literally bombarded with departures from a tenuously established "new wisdom" that presents us with more and more still "newer" wisdoms. In thermodynamics and throughout the range of the chemistry underlying both life and nonlife, for example, Ilya Prigogine has recast a world of chaos and order into a new worldview of order within, out of, and through, chaos. In brain research, Karl Pribram has scrambled our understanding of how the brain operates with the mysterious mathematical mix of holographic and holonomic theory. In physics, David Bohm has pushed the "mystical" strain already evident within Bohr to the ultimate with his view of the mighty interaction of a timeless, implicate order with our ever-present explicate order. And Rupert Sheldrake has jolted biology—and indeed the science of communication in its broadest sense—with his argument for the mass-memory storage of cumulating and eternal morphogenetic fields.

When each of these new views first appeared we were warned away from them by the guardians of prevailing scientific faiths. But from the beginning there was something about these "heretics" that felt intuitively right to many thousands of readers—scientists and nonscientists alike. However off the mark they might prove to be as science progressed, one sensed an underlying great new body of "truth" not only struggling to emerge but also—because of its capacity for a long-overdue transformation of mind—most insistent on its birth.

The thought of these new scientific explorers—and of many more, ranging across all fields of modern science—is the new body of "truth" to which Ervin Laszlo brings his well-known skill as a magnificent synthesizer of vital knowledge. One problem with this new knowledge is that, as it is fragmentary, we can do no more than guess at a meaningful whole. And because we lack this sense of a whole to which we can relate the fragments, we tend to quickly lose the insight they convey. Another problem is closely related—this newest development of the "new science" comes to us couched in the language of the fragments: the specialist languages and concepts of highly complex and often bewilderingly multidisciplinary fields. We pretend to understand, but if we are honest about it, practically all of us who try to keep up with this bombardment must confess to difficulties.

In its rare power to bind the glimpses of a new world together in a meaningful and accessible whole, many readers will find Laszlo's *The Whispering Pond* a most compelling intellectual journey. It is a forceful model of how to present the kind of bold, breakthrough thought that is needed in our time. Its discipline, scope and power of reason call to mind the strength of the past ages of great philosophers, only this time we have the added power of Laszlo's incisive grasp of the latest developments in the sciences. But *The Whispering Pond* is considerably more than a much-needed synthesis of prior thought: as we are being led into the new territories of science, what we are reading gradually becomes, in effect, the journal of an original explorer while in the process of discovery. The reader is getting in on the ground floor of thought as it is happening, chronicled by a science writer some years ahead of his time.

The new vision of science emerges through the way Laszlo develops his "quasi-total vision" out of a wedding of the "super-theories" of the new physics (cosmic inflation, black holes, superstrings, etc.) with the new theories of the evolution of all life, including our own. Something of the grandeur of this new vision comes to the fore when we view it as an account of emergence, of evolution, of nature taking off from physical reality and attaining to the ethereal spheres of life, mind, and consciousness. This evolving whole-field universe registers and feeds back all that we do and all that we think—we are not just immersed in it, we are part of it. To our particularized intellect it may appear separate, but even as separate it fills us with wonder. It is a universe of creative connection, of evolution based on information and memory.

In this new development science becomes an instrument for the recovery of the wholeness of the universe, and with it the wholeness of all that exists in it, including ourselves, our thoughts, our feelings, our dreams, our fears and our hopes—and, above all, our visions and our creativity.

Laszlo's power as both synthesizer and innovator is evident in the way he is able to single out the key unsolved mysteries in each of a number of scientific fields. One by one, with fascinating ingenuity, he shows us how a key concept—and behind it the emerging quasi-total vision—provides a holistic rather than a fragmentary answer for each of these cognitive "black holes" in our understanding of mind and universe.

I would particularly like to call attention to the cosmological implications outlined by Laszlo—a *tour de force* of great hope for humanity in the long term. Rather than one Big Bang at the start followed by a gradual running down, Laszlo describes a continual recycling of sequential universes, each linked to the other by the memory storage of an interconnecting field. Here we also find what may be the single most productive statement in many years on two matters of exceptionally wide interest. One is the question of how so-called psi phenomena (ranging from telepathy to past-life therapy), hitherto excommunicated by formal science, fit into the scheme of life. The other is the nature of the bridge between Eastern and Western thought, between spiritual insight and experimental science.

The Whispering Pond is an enormous contribution to our understanding at a critical time in human evolution. It gives us the vital new fragments of emergent "truth" in language we can understand. And it provides the even more vital sense of the meaningful whole into which these fragments fit, which we have been lacking. This book, and the pioneering scientific study it is based on—Laszlo's *The Interconnected Universe*—call to mind that watershed statement of the 18th century, *The Critique of Pure Reason*. There a philosopher with a similarly amazing capacity for integration—Immanuel Kant—so transcended in his synthesis the science and philosophy of his time as to establish a new framework for practically all of modern thought. It will be interesting to see if history repeats itself.

FOREWORD

II

KARAN SINGH

President of the International Centre for
Science, Culture and Consciousness,
Jammu and New Delhi, India

ERHAPS the most significant development in recent times which, though the subject of several important books, has still not received the attention it deserves, is the growing convergence between the mystical worldview (predominantly, but by no means exclusively, Eastern) and the emerging paradigm of reality among scientists at the cutting edge of contemporary knowledge. *The Whispering Pond,* the latest in Ervin Laszlo's important series mapping out the geography of reality, makes this point, and does much to rectify it.

With astonishing incisiveness and clarity, *The Whispering Pond* propounds a breathtaking vision. Its most significant upshot is that the scenarios of cosmic fate are likely to be open; fate and destiny are not sealed, and the future may not only happen but could also be created. This is strikingly similar to the hypothesis put forward by Sri Aurobindo, the greatest Hindu philosopher of modern times, that with the advent of Man we have for the first time a creature that can, and must, actively cooperate with the evolutionary force

xiii

in order to accelerate its processes. The next step in human evolution will not be in the outer configurations of the body, but in the
inner constellations of consciousness. The Hindu concept of Yoga
—the linking of human and cosmic consciousness through psycho-
physical practices—provides a methodology for this creative
transition. In the West, the works of C.G. Jung and Teilhard de
Chardin—to mention but two outstanding thinkers—represent
similar attempts to probe the inner dimensions of the evolution of
consciousness.

In the light of the globalization of human civilization taking
place before our very eyes, the evolution of a global consciousness
is urgently needed if mankind is not to destroy itself and all life on
this planet by its inability to responsibly manage its technological
ingenuity. For such a global consciousness to arise, a worldview in
which science and spirituality converge is a necessary development. The publication of *The Whispering Pond* is a significant step
in this direction.

The image, outlined in the concluding chapter, of the cosmic
dance with a subtle and basic universal energy field raises interesting parallels and equivalences with Hindu thought. As Laszlo
writes, the possibility that the world consists of more than blind
flows of energy and randomly appearing and disappearing configurations of matter cannot be disregarded. Indeed, the Upanishadic
worldview, which represents one of the earliest and certainly
the most comprehensive articulation of this viewpoint, starts precisely with the concept that behind all the constantly changing
formations of matter/energy—whether intergalactic or subnuclear
—stands the eternal and unchanging Brahman, the immutable
"field" upon which all manifestation takes place. It is, as it were, the
eternal screen upon which the phantasmagoria of manifestation is
projected in the great unending cycles of time.

Then there is the concept of *Akasha,* the element which
records permanently all that ever occurs anywhere in the cosmos
through a process that is analogous to that which occurs in what

Laszlo names the Ψ field. In the *Shvetashwatara Upanishad,* Shiva is described both as the "creator of time" and the "destroyer of time." In other words, while manifestation occurs only in a time-space continuum, the supreme reality, being eternal, predates and outlives the periodic movement into cosmic manifestation.

It is the image of Shiva Nataraja, Lord of the Cosmic Dance, which symbolizes better than any other human artifact the new dimensions of cosmology. In one hand Shiva holds the drum—the creative word—through which millions of galaxies spring into existence every second; in another the fire that symbolizes the destruction of these worlds. The universe is viewed as *samsara*—that which is in constant change. The other two hands—one raised in benediction and the other pointing to his raised foot—provide a path of personal contact with the divine within the cosmic cycle of creation and destruction.

Our universe is currently estimated to have been born 15 or perhaps only 8 billion years ago. But how can we assume that this is the only universe that exists? Hindu cosmology would tend to support an infinite number of universes, all drawing their existence from the all-pervasive Brahman. Regardless of the number of Big Bangs, the Brahman retains its infinite plenitude. As the Vedic hymn says, "This is infinite, that is infinite, the infinite springs from the infinite. If the infinite is taken from the infinite, the infinite still remains." On this score, the scope for creative cross-cultural analysis is endless; a full-length exploration of the parallels between Hindu cosmology and modern scientific concepts is clearly indicated. The emerging scientific vision of the universe can be elaborated and illuminated by a penetrating interlocution of the Upanishadic model. Laszlo is open to this exercise, which is why his luminous work not only elicits admiration, but arouses a creative response.

What is particularly striking about the exposition of *The Whispering Pond* is its elegant simplicity and—mercifully for the lay reader—its entirely nonmathematical presentation. Laszlo's extensive experience in interacting with a wide spectrum of creative individuals and institutions—scientific as well as educational—has given him an unusual capacity to communicate his insight into the nature of the universe we inhabit, and his ability to move away from the rigidities of the dualistic/materialistic paradigm (what I call the Cartesian–Newtonian–Marxist model) enables him to look deeper into the new concepts of science in order to understand why complex and consistently ordered phenomena emerge in the vast trajectories of evolution, rather than no order and consistency at all.

Among billions of galaxies one is ours, among billions of stars in this galaxy one is ours, among billions of living beings on this planet one of them is ourselves, but such is the grandeur and mystery of being human that we can move toward a comprehension of the unutterable mystery of existence. We who are children of the past and the future, of the Earth and the heavens, of light and of darkness, of the human and the divine, of the world and beyond it, at once evanescent and eternal, within time and in eternity, have been, incredibly, endowed with enough capacity to comprehend our condition, to rise above our terrestrial limitations and, ultimately, to approach the possibility of transcending the throbbing abyss of space and time itself. This is the unprecedented adventure that awaits those who venture, with *The Whispering Pond,* into the uncharted regions of the new consciousness, illuminated by the worldview emerging at the cutting edge of the contemporary sciences.

INTRODUCTION

N 1597, almost a century before Newton, the English philosopher Francis Bacon wrote that knowledge is power. His dictum applies today more than ever, with the proviso that the knowledge in question either comes from the sciences, or has some measure of scientific legitimacy. Because science has become a major—perhaps *the* major—force shaping today's world.

Whether we like it or not, science has turned into a kind of secular religion. While in the Middle Ages it was the Church that was linked with the State in a master-servant relationship, in the three centuries since the discoveries of Newton the role of master has been taken over by the apostles of science. The establishment of science has acquired an educated priesthood with privileged access to esoteric knowledge. This coterie legitimizes public policy and sets standards of behavior. Its sacred scriptures are the treatises of theoretical and experimental scientists. The tenets of physicists justify engineering in both the natural and social sciences; the discoveries of biologists influence legislation on matters of health and well-being; the formulas of microeconomists are guideposts for the management of individual enterprises; and the doctrines of macroeconomists influence the management of national and international economic processes.

The long-term evolution, and even the sudden revolutions, of contemporary societies are driven more by the social and techno-logical spin-offs of scientific innovations than by the power and will of politicians and managers. The breakthroughs of microelec-tronics have opened the information superhighway to global traffic and now bring to the fingertips of those who navigate it ideas and images from every conceivable field of interest—from local gossip to global crises. The technical applications of information and con-trol innovations enable many people to enjoy reduced working hours and increased leisure-time, and to talk with anyone instantly and often at negligible cost. Innovations in transport technologies enable massive streams of tourists and business people to travel anywhere on the six continents in a matter of hours in considerable comfort and safety. Breakthroughs in biotechnologies make pos-sible the enlargement of the food supply and the extension of the human life span, with fresh cures for the many diseases that still inflict the human condition. Paradoxically, even the absence of all-out war is due in some measure to advances in science-based tech-nology: modern weapons have become so powerful that they now endanger the potential victors themselves, and reduce the spoils of war to heaps of rubble—that may in addition be poisoned or radio-active.

To the list of science's achievements and benefits we could also append a list of its shortcomings and disadvantages. The short-sighted applications of science depress the quality of the environ-ment and overexploit precious natural resources, and they polarize societies into separate groups: those who can cope with science's complexities and those who cannot—or will not. And, at least on first acquaintance, scientific tenets convey a dehumanized picture of the world—dry and abstract, reduced to numbers and formulas without feeling and value, heart or soul.

Whether we admire science or fear it, whether we adopt its implications and applications or reject them, we must acknowledge that science infuses our lives and penetrates our ways of thinking in more ways than most of us realize—and some of us wish. The elements of science that enter our lives are not just its technical applications, but such "soft" factors as our view of nature, man and world. The concepts produced by the sciences shape our perceptions, color our feelings, and impact on our assessment of individual worth and social merit. They enter into the bundle of ideas, emotions, values, and ambitions that we call human consciousness—the warp and woof of our immediate experience.

The question is no longer whether science affects our lives and our consciousness, but whether it affects them for the better or for the worse—whether it helps us meet our goals and realize our dreams, or exposes us to inhuman conditions with attendant shocks and surprises.

The "scientific worldview" that affects most people's minds is not a happy one. In this view the essential features of the human species are the result of a succession of random, accidental events in the history of life on Earth, while the unique features of the human individual derive from the fortuitous combination of genes with which he or she was born. The ongoing struggle for survival in which every individual, every enterprise and every society is ceaselessly engaged has made us into bundles of egotism, separate from all that lies beyond the limits of our body and the compass of our personal and professional interests.

But this is not the view of the world warranted by the concepts and theories of the contemporary empirical sciences. Beyond the accidents of mutation and natural selection in a world dominated by the random concourse of atoms and particles, leading-edge science is discovering a deeper logic. This does not mean that scientists are appealing to a transcendental mind or spirit to guide the processes that have led to the human species; rather, they are discovering the integral dynamic of the processes that have brought human beings (and all things in the observable universe) on the scene. In the embracing vision that is now emerging, everything that has evolved in the universe—Mozart and Einstein, you and me, the greatest of galaxies and the humblest of insects—is the result of a stupendous process of open-ended yet nonrandom self-creation. Nothing that has ever evolved exists separately from all the rest: all things are connected, all are part of an organic totality.

In the emerging vision of leading edge science the world is a seamless whole composed of its parts. More than that, it is a whole in which all parts are constantly in *touch* with each other. There is constant and intimate contact among the things that coexist and coevolve in the universe; a sharing of bonds and messages that makes reality into a stupendous network of interaction and communication: a subtle but ever-present whispering pond.

At a time when we and our societies are becoming drawn into an interacting and interdependent web of technology, finance, production, consumption, and even leisure and culture, it is vital that our consciousness be infused with this new vision, rather than with the old. We need to realize that enduring connections among humans, and between humans and nature, is an extremely important, as well as intensely meaningful concept. It is this kind of insight that could re-establish harmony and balance in a world of vulnerable interdependence and ever more chaos.

The vision of a universe of subtle and constant connections is a trustworthy beacon to light our way, as we tread the individual path of our personal fulfillment within the shared path that will decide the future of our species. This is the conviction that has led to the writing of this book, and it is this insight, the author hopes, that will emerge on reading it.

A NOTE ON THE CONTENTS AND ORGANIZATION OF THIS BOOK

The technical details of the vision of the world that results from the current developments in science are the subject of a separate treatise by the author, intended primarily for the scientific community (*The Interconnected Universe—Conceptual Foundations for Transdisciplinary Unified Theory,* World Scientific Limited, London, New Jersey and Singapore, 1995). The present book bypasses the rigors of technical exposition—it comes right down to the basic questions of meaning and coherence, and of human significance.

Part One describes the "established" worldview that underlies the mainstream theories and concepts of the natural sciences, reviewing what most scientists believe we already know of cosmos, matter, life, and mind. This vision is accomplished as never before, and it is important in its own right. It is not the final word, however: the current vision is as yet incomplete, harboring numerous fuzzy areas and even some black holes.

Part Two focuses on these blurs in the established image, surveying each major domain of nature in regard to not what science claims to know about it, but what remains puzzling and paradoxical in its established theories.

Part Three traces the latest developments at the leading edge of scientific research. Science is an open enterprise: when faced with puzzles and paradoxes, scientists set out on fresh paths of inquiry, leaving behind the established concepts and theories. Such a "revolution" is under way today. It unfolds in physics in the domain of GUTs (the Grand Unified Theories that seek a unitary description of the physical universe); in biology, where the holistic concepts of developmentalism challenge the time-honored doctrines of Darwinism; as well as in a transdisciplinary domain that cuts across the classical fields and seeks an understanding of how life has evolved out of non-life, and mind out of life.

The concluding **Fourth Part** of this book goes further. Here we anticipate the vision science will produce at the dawn of the 21st century. Our forecast of the coming scientific revolution is based on recent work in the physical and the life sciences, as well as on the avant-garde theories that seek an understanding of the evolution of life and mind from the physical universe since its fiery origins in the Big Bang—or perhaps before. Our review leads to the crowning image of these explorations: the vision of a memory-filled, interconnected and self-creating cosmos—a fathomless and timeless whispering pond.

The intellectual journey that has led the author to the shores of this pond has been fascinating and fraught with intensity and excitement. The journey that awaits the reader is likely to be no less fascinating, and its excitement will depend on the questions he or she will dare to ask and the courage with which he or she will follow up the answers that will present themselves.

For the author it only remains to wish the reader . . . *bon voyage!*

PART ONE

THE ESTABLISHED VISION

CHAPTER 1

THE EVOLUTION OF THE COSMOS

SETTING OUT

A S IN ANY real-world voyage of discovery, when setting out on
an exploration of the emerging vision of science we must start
with the near shore: what we already possess in the way of
scientific knowledge. Of course, we shall not linger unnecessarily
at that shore since the established body of knowledge is not permanent: here once and for all. The near shore, with its definite
contours and occasionally exquisite detail of understanding, must
be sooner or later left behind for open seas and as yet hazy
horizons. This is because the hallmark of any item of scientific
knowledge—the substance of the near shore—is that, though it
may be currently verified (more cautious scientists would only say
"confirmed"), it can always be "falsified." Indeed, falsifiability, as
philosopher of science Sir Karl Popper pointed out, is the principal
feature of any concept or theory that refers to the scientist's world
of observation and experiment.

As scientists and historians of science well know, with the
advance of knowledge not only are more facts and items added to
the received corpus of scientific knowledge; also some items in that
corpus turn out to be in error. They are falsified by the light of new
evidence, or by new and more consistent ways of construing the
evidence. Occasionally, even fundamental assumptions about the
nature of the observed world become exposed to doubt. Of course,

basic assumptions are not definitely discarded on the basis of evidence alone: before undertaking such a radical step, scientists cast about for alternative assumptions and workable hypotheses. Only when solid anchoring insights have been found are the previously held basic assumptions abandoned. Then, however, science undergoes a revolution; a so-called "paradigm-shift." Scientists pick up their equipment and transfer from the previous shore of familiar knowledge to new, recently conquered continents.

By and large, this is how science makes progress: through periods of accumulation alternating with periods of revolution. In the course of this progression ever new horizons are glimpsed by leading edge theoreticians, and when they succeed in describing them, their experimentalist and mainstream colleagues undertake the detailed explorations. The previous shore of received knowledge fades into the past; the new and hitherto hazy shore advances ever more sharply into focus. When it is sufficiently mapped, scientists find themselves with a new and different view of their segment of the empirically known world.

Such a change of scenery is of more than academic interest. Scientists not only observe; they also interact with what they observe. They do so already in testing their observations, and then, with the help of their technologist and engineering colleagues, in finding ways in which the observations can be put to practical use. Evidently, not every area of observation allows such use—explorations of distant galaxies, for example, would hardly qualify in this regard. Yet in some way almost every fundamental assumption about the nature of reality has implications for human life—humanity, after all, is part and parcel of the reality mapped by scientists. And that means that familiarizing ourselves with the new horizons discovered and explored by scientists is of both theoretical interest and of practical value. It gives us the best insight we currently possess into the nature of scientifically knowable reality—and it also gives us an effective handle to interact with some aspects of that reality.

To outline and explore the latest horizons discovered at the leading edge of the contemporary sciences is the ultimate objective of the voyage we undertake in this book. We cannot, of course, leap directly to the new continents; first we have to cover the intervening stretch of seas, even if they are hazy or turbulent. We must launch our craft from the near shore: the shore of received knowledge. Familiarizing ourselves with the principal landmarks of this shore is the task of Part One of these explorations.

The most general contours of the near shore of received knowledge are conveyed by those branches of the empirical sciences that investigate the basic laws and regularities of the universe. These are the *physical cosmologies*. On their domain scientific curiosity coincides with public wonderment. Because, for hundreds if not thousands of years, few questions of scientific interest have excited the popular imagination as much as the nature of the cosmos as a whole. What lies beyond the familiar world of trees, rocks, rivers and seas that surrounds our cities and our abodes? The stars seem equally remote when viewed from the ground or the highest mountain top. Were they always there, or have they been created? And if the latter, was their creation due to the same forces that created the trees and the rocks — and us human beings?

Such questions are now joined by new ones. Did the universe really begin in an explosion called the Big Bang—and if so, what was there before? Can life continue indefinitely in the vast reaches of cosmic space, or will it—must it—come ultimately to an end? What about the stars and planets, and the countless galaxies themselves? Is there life on some of them? And what is their future, and the future of the space and time in which they subsist?

These are the largest questions we can ask of science and still hope to get a reasoned answer. That such an answer is forthcoming to some (though not necessarily all) of these questions is a sign of remarkable achievements in the contemporary physical sciences. Notwithstanding competition between various *cosmologies* (theories of the nature and development of the cosmos) and *cosmogonies* (theories of the origins of the cosmos), scientists are now in overall agreement on the broad outlines of the origins and evolution of the universe.

This was not always the case. Though cosmologies and cosmogonies are as old as human intellectual history—ever since *Homo* appeared on this planet, the mystery of a jet-black sky dotted with twinkling stars has attracted his attention and challenged his imagination—previous conceptions were metaphysical and speculative, if not downright esoteric. It is only in the last 200 years that any theory regarding the origins and development of the universe could qualify as scientific.

A HISTORY OF SPECULATION

The ancient Sumerians, Babylonians, and Egyptians, as well as the Indians and Chinese, produced detailed accounts of what they believed to be the ultimate nature of man and cosmos. Mythic cosmologies were also developed in the pre-Columbian civilizations of the Mayas, the Incas, and the Aztecs, as well as by tribal cultures in Africa. Their accounts were myths of creation that deduced the existing world from supernatural entities wielding supernatural powers. Sometimes these entities were seen to be in conflict with one another, and the nature of the manifest world symbolized the victory of one over the other. At other times the opposing forces were looked upon as creative, producing the tension—the yin and the yang—out of which the varied spectacle of the known universe has sprung forth. In most cosmologies, especially in the East, the process that gave rise to the world involved many stages, with one stage giving rise to the next.

When the thinkers of classical Greece relinquished mythical accounts in favor of rational speculation, more elaborate cosmologies appeared. While the details of the theories were as varied as the thinkers who expounded them, they had common basic features. The genesis of the world was deduced from the fewest possible elements or basic principles, such as water, earth, fire, or air, or some combination of them. The process itself embodied a progression from the less to the more perfect—contrary to the "golden age" myths of still earlier times, where the world-process was an inexorable descent from a prior era of perfection.

The Greek philosophers were largely agreed that the universe has no natural limits either in space or in time, and is governed by a set of unalterable laws with their own recurring rhythm. When they assumed that the deeper reality beneath the diversity of the sights and sounds that reach the eye and the ear is coherent and unitary they echoed an ancient insight of Eastern philosophies: that all there is in the world emerged stepwise from an original Source. That Source itself is indivisible, and spaceless and timeless in its primordial essence. But, unlike the sages of the East, the Greeks insisted that this original Source and its gradual diversification into the observed world could be grasped without recourse to mysticism.

In the golden era of Greek philosophy, the cosmic process was both optimistic and rational. The world progressed from "chaos" to "cosmos" under the rule of an ordering principle. For Plato this principle was a principle of intelligence (*nous*), while for Aristotle nature itself was the cause of order.

These views, though they were influential for many centuries, underwent important modification with the rise of Christianity in the West: the originating creative source became identified with God, the omnipotent Creator of the Earth below and the sky above. The stars themselves were not independent entities but fixed appendages to a huge sphere that was the backdrop of the human world; it rotated around the Earth once each day. The Earth occupied the exalted position at the center of the universe, being the immovable abode of man, who was God's creation in His own likeness. The universe was unchanging and indeed unchangeable: it existed for all times as God had created it. But in itself it was finite in space, bounded by the rotating stars. The infinite space beyond it was filled by the infinite God.

In the 16th century, the astronomer Tycho Brahe did away with the crystal spheres that the medieval mind placed behind the Sun, the moon and the stars to explain their motion. In Brahe's system the Sun revolves around the Earth, and the planets around the Sun. This refined but still geocentric cosmology received a major (though intentionally cautious) blow when Nicolaus Copernicus announced, first in 1543, that the computations of astronomers could be made simpler if it was assumed that it was the Sun that was at the center of the universe and not the Earth. And, offered Copernicus, this could be true, since nature loves simplicity. In the Copernican heliocentric universe it is the Earth that rotates on its axis once every 24 hours and thus produces the alternation of day and night, and the apparent rotation of the night sky.

The Copernican revolution had another significant consequence: there was no longer a need to assign the stars to a crystal sphere around the Earth. They could be located anywhere, at any distance, and they could remain fixed in their position. This accorded with a view taken in 1440, a full century earlier, by the German Cardinal Nicholas Krebs of Cusa. In his *De docta ignorantia* he proposed that the universe is infinite both in extension and in its constituent elements. The Earth has the same status and dignity as all the other stars; like God, the center of the universe is anywhere and everywhere. Following the publication of the Copernican theory, this was the view that was embraced by Giordano Bruno, the visionary scientist who ended up being burned at the stake for his "heresies."

The wider acceptance of the Copernican view was due to the English mathematician Thomas Digges, who in the year 1576 included a description of the heliocentric system in a book in which he translated major parts of the Copernican treatise. Copernicus himself did not reject the doctrine of fixed stars at a finite distance from the Earth (they could not be at *infinite* distances because then, as Aristotle had already pointed out, their rotation would have required infinite speed), but Digges did. He affirmed categorically that the universe was infinite. The distant stars were themselves suns and were dim only because of the great distance that separated them from the Earth. Stars like the Sun, he said, are located in the universe at distances stretching out to infinity.

Once the infinite universe concept was accepted, the question that occupied the minds of scientists was whether stars continued infinitely in the universe the way we observe them in our own region, or thinned out ultimately to nothingness. Newton was of the opinion, published in 1692, that the universe must consist of many individual masses separated by great distances. Matter tended to clump together under the force of gravitation, and these giant aggregates were distributed throughout the infinite space of the universe. That we could see only a finite set of stars was due merely to the limitations of our telescope. Immanuel Kant elaborated the notion of an infinite universe made up of distinct "island universes"—this view prevailed from the mid-18th century onward. It was reinforced when, by the early 19th century, William Herschel mapped our galaxy (which he called "nebula") in remarkable detail and plotted other galaxies as well. The latter, he said, represent island universes beyond our own.

The next major development in scientific cosmology was due to Einstein. He published his model of the universe in 1917, a year after the general theory of relativity, on which it was based. Einstein's model disagreed with the concept of a vanishing series of "finite islands in an infinite ocean of space." Energy from the stars at the more populous center, he pointed out, would be continually leaking to the thinly populated regions in the form of radiation, and it would be swallowed up in the infinite reaches of space. Einstein also rejected that, in the course of infinite time, entire stars would be knocked by chance collisions into the surrounding space, leaving the observed universe to depopulate, ultimately to zero. Rather, he joined time with space and observed that the non-Euclidean—curved and four-dimensional—geometry of this continuum is finite though it is unbounded. Space-time curves back on itself, so that a space traveler who went far enough and for long enough would eventually return to his point of origin—though to him it would appear that he had traveled in a straight line.

In Einstein's mathematical cosmology matter is treated as if it were spread homogeneously throughout space-time. Because matter—in the form of mass—is subject to the law of gravitation, in this universe matter would tend to clump into a single mass at the center. Since this is not the case, Einstein introduced a repulsive force (the so-called cosmological constant) which would precisely balance the attractive force of gravitation. This, he said, keeps the universe in a steady state forever.

Einstein's stable three-dimensional universe within its infinite four-dimensional space-time had pleasing mathematical properties; it even had a definite and plausible world radius (estimated at 109 light years—almost the same as the range of the 200-inch telescope at Mt. Palomar). Nevertheless, Einstein's steady-state model had to be surrendered. In 1917, the Dutch astronomer Willem de Sitter found another solution to Einstein's relativistic equations. De Sitter's solution indicated that when matter is introduced into the space-time continuum, it acquires a velocity away from the observer, and this velocity increases with distance. Parallel to this, as distance from the observer increases, time slows down, coming to a stop at the limit of observation.

It was not long before the English astronomer Sir Arthur Eddington realized that in Einstein's universe any expansion or contraction of matter would result in continuous motion in the direction it was initiated. In consequence Einstein's universe appeared to furnish but a transitory phase, leading (if matter's motion was that of expansion) to de Sitter's universe. The mathematics of this unstable universe were discovered by the Russian mathematician Alexander Friedmann in 1922. His solution modified Einstein's cosmological constant and introduced a constant that could be positive, negative, or zero. According to the value chosen, the universe is seen as expanding, contracting, or moving ultimately toward zero-expansion, that is, into a steady state.

THE RISE OF BIG BANG COSMOLOGY

Starting in 1923, U.S. astronomer Edwin Hubble carried out a convincing series of demonstrations at the Mt. Wilson Observatory that demonstrated the astronomical version of the Doppler effect (the effect of frequency differences between waves emitted by sources approaching and receding: those emitted by approaching sources are compressed to higher frequencies, while those coming from receding sources drop to lower ones). Distant galaxies exhibited the lower-frequency "redshift" typical of receding light-sources, and the more distant the galaxies were, the greater was the shift.

With this explanation, the expansion of the universe appeared to be definitively established. The big question that remained was, how did the cosmic expansion get under way? Though accomplished steady-state cosmologies were available by mid-century (following a suggestion of Sir James Jeans, they replaced the theory of matter and energy leaking into infinite space with the theory of continuous creation of matter-energy in the central regions), the way was open to the modern theory of the Big Bang.

In its currently discussed form, the mainstream "Big Bang scenario" dates from the 1980s. Its postulates have been confirmed by the computer analysis of about 300 million observations made over the course of the year 1991 by Nasa's Cosmic Background Explorer Satellite (Cobe). Cobe's detailed measurements of the cosmic background radiation show that variations in this radiation field are genuine fluctuations deriving from the Big Bang, rather than—as was sometimes suspected—distortions caused by radiation from astronomical bodies. The variations date back to a time when the universe was about 300,000 years old: they imply the existence of huge clouds of matter, the precursors of galaxies. These are thought to be due to minute fluctuations that occurred in the dispersion of the cosmic fireball less than one trillionth of a second after it exploded.

The Big Bang itself is believed to have consisted of two phase-changes in rapid sequence. The first led to the explosive inflation of the fluctuating "vacuum" that is the cosmic womb from which the universe itself has emerged. This phase follows de Sitter's equations and is known as the de Sitter universe. In the second phase the inflationary universe transformed into the more sedately expanding Robertson–Walker universe which is the one in which we live today. A further phase change came about subsequently, when the universe was between 50,000 and 1 million years old: matter uncoupled from radiation. Space became transparent, and particles of matter established themselves in the expanding reaches of cosmic space. From this time onward, the history of the known universe has been the history of the evolution of galaxies and stars in space and time.

According to the current view, the matter that now populates the vast reaches of cosmic space was synthesized in the first few milliseconds after the Big Bang. But matter did not emerge in space and time all at once and complete in every detail, like Venus from the sea. At the extremely high temperatures that prevailed in the very early universe there was only superheated plasma: atoms did not exist, since thermal noise had prevented electrons from associating with nuclei. Then, as the plasma cooled, electrons began to circulate around nuclei and a gas of atoms emerged. At that time galaxies condensed out of the plasma, and stars condensed within the galaxies. With further cooling, various atoms configured into molecules. Still further cooling allowed the formation of complex molecules, transforming matter from the gaseous into the liquid state, and then into the familiar solid crystalline form.

As matter aggregated under the force of gravitation, galaxies formed, and within galaxies, stars, and stellar systems. On suitably located planets of active stars molecular and crystalline structures may have configured further. Cell-like structures—so-called protobionts—could have emerged and, provided thermal and chemical conditions were favorable, this could have opened the door to the evolution of the still higher-order configurations that underlie the phenomena of life.

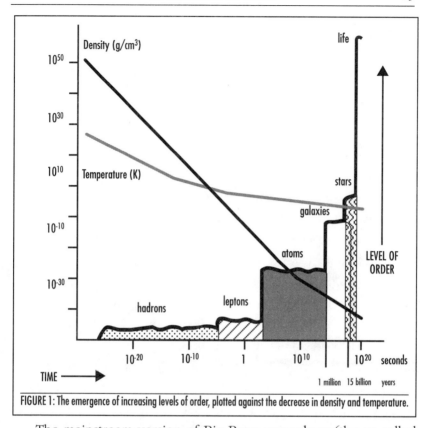

FIGURE 1: The emergence of increasing levels of order, plotted against the decrease in density and temperature.

The mainstream version of Big Bang cosmology (the so-called "standard BB-scenario") assigns a specific sequence of events to this process with a corresponding time-frame. The first particles to be synthesized were the hadrons (heavy particles such as protons and neutrons): they were created within 10^{-24} to 10^{-3} seconds of the Big Bang—that is, at a time when the universe was far less than one thousandth of a second old. They must have existed as free unbound entities, but in the extreme density of the early universe must have frequently collided and interacted with each other. The extremely high temperatures of this epoch—estimated at around 10^{15} degrees on the Kelvin scale—must have prevented the particles from combining into atoms. The hadrons had most likely self-annihilated in the process, decaying into photons and adding to the intense fireball of radiation. *After* the first millisecond (10^{-3} seconds) the fireball cooled beyond the threshold where hadron creation was possible, allowing particles such as electrons and neutrinos (lighter particles known collectively as leptons) to achieve dominance. The expanding universe became less dense, its matter-content dropping from 10^{30} to about 10^{10} grams per cubic centimeter. But, after the first second had passed in the life of the universe, the leptons too had self-annihilated into photons, further fueling the fireball with high-frequency radiation.

In the era of the first second, the photons far outnumbered the particles of matter: the energy of the universe was mainly in radiation. The existing matter-particles could not aggregate into higher clusters; the intense radiation field broke up all further configurations. Matter existed only as a thin precipitate in an intense radiation field.

When the universe reached the respectable age of 100 seconds, the average temperature dropped to about 10^5 K and average density to 10^{-10} g/cm^3—values much like those that still exist in the interior of active stars. This permitted the electromagnetic clustering of hadrons and leptons into neutral atoms. Hydrogen, where a single electron is bound electromagnetically to a single proton, was the first element to emerge. Since the "cooking temperature" of the remaining fireball was still high enough to have fused two hydrogen atoms via the proton-proton cycle into an atom of helium (at the presumed rate of one helium atom for every ten hydrogen atoms), the young universe became filled with a gas of hydrogen and helium. Then, as matter built up sufficiently to separate off from radiation, the epoch of galaxy formation dawned.

The time-frame of galaxy formation is still a matter of controversy; various models vie with each other. It appears likely, however, that galaxies would have formed in a universe that was somewhere between 10^6 and 10^9 (1 million and 1 billion) years old. During that epoch the average temperature had dropped into the region of 300 K, and density had decreased by a further factor of 10 billion to about 10^{-20} g/cm^3.

Within the vast galactic clouds of hydrogen and helium, the uneven distribution of hydrogen and helium particles produced further gravitational clumping, heating up the aggregates of matter to the point where nuclear cooking temperatures were reached again—this time within newly forming stars. This led to the synthesis of some of the heavy elements, such as carbon, oxygen, and iron.

The basic nuclear transmutation process, from hydrogen to helium, produced a constant stream of radiation from active stars into surrounding space. Where the stars had planets circling them, their satellites received a part of the energy flow. Provided they were at distances where the flow was neither so hot as to boil water, nor so cold as to freeze it, more complex aggregations could have come about in the already complex mix of elements. Here and there supramolecular configurations must have been produced, and some—like those on this planet—could have reached the level of complexity where the self-maintaining metabolic processes associated with life could get under way.

According to the mainline theory, the universe today is about 15 billion years old (though it may also be that it is only 8 or just 7 billion years "young"), with an average density of less than 10^{-30} g/cm^3 and a background temperature of 2.7 K. Our Sun is one of more than 200 billion stars in our own galaxy and of about 10^{22} (10 billion trillion) stars in the universe. Our galaxy, in turn, is one of many in the Local Group, beyond which there are about 100 billion more galaxies, some of unexpectedly large dimensions.

THE COSMIC SCENARIOS

This, as far as we know, is how the universe is today. But how will it be tomorrow . . . and in the more distant future?

To this question various answers are possible. The universe may be open (infinitely expanding in cosmic space); it may be closed (reversing back on itself in a final Big Crunch); or it could be in a steady state, balanced on the razor's edge between expansion and contraction. If it is flat, it will reach a steady state where the outward-pulling inertial force of the initial explosion is precisely balanced by the inward-pulling force of gravitation, and it will remain in that state forever. Thus a flat universe, though it is finite in space (it has a spatial boundary beyond which it will not expand), is infinite in time. However, if the universe is open, the force of expansion carries galactic matter further and further apart in space: the open universe is infinite in space as well as time. But, if the force of gravitation exceeds the force of expansion the universe is closed, and then it will stop expanding (perhaps around 1,000 billion years after the Big Bang) and thereafter begin to contract at ever higher velocities. It will collapse back on itself in the Big Crunch at a time horizon estimated at around 2,000 billion years. In consequence, the closed universe is finite both in space and in time.

We do not know at this time whether the universe is open, closed, or flat. This depends on the exact amount of matter in cosmic space. If matter *exceeds* the critical threshold of $5 \times 10^{-27}\,\text{kg/cm}^3$, we are living in a closed universe. If it is just *at* that threshold, the universe is flat; if it is *below* it, the cosmos is open.

But as far as the ultimate fate of matter and life is concerned, the alternatives make little difference. In any case, the constructive phase in the evolution of the universe cannot continue indefinitely: cosmic evolution must sooner or later reverse into devolution. The reversal will come at different times in different places, but when it comes, it will be irreversible. Ultimately all matter in the cosmos will degrade and disappear.

The macrostructures of the universe—stars, stellar systems, galaxies, and galactic clusters—will decay as well. The cosmic downhill scenario can be reconstructed in its essentials as follows:

About 10^{12} (1 trillion) years from now, no more stars form. The existing stars have already converted their hydrogen into helium, the main fuel of the supercompacted but still luminous white-dwarf state. Then also helium is exhausted, and galaxies assume a reddish tint. As their stars cool still further, the galaxies fade from sight altogether.

As energy is lost in galaxies through gravitational radiation, individual stars move closer together. The chance of collision among them increases, and the collisions that do occur precipitate some stars towards the center of their galaxies and expel others into extragalactic space. As a result, the galaxies themselves diminish in size. Also the galactic clusters shrink. At last both galaxies and galactic clusters implode into black holes.

At the staggering time-horizon of 10^{34} years, matter in the cosmos is reduced to radiation, positronium (pairs of positrons and electrons), and compacted nuclei in black holes. Black holes themselves decay in a process described by Stephen Hawking as "evaporation". A black hole resulting from the collapse of a galaxy evaporates in some 10^{99} years, and a giant black hole containing the mass of a galactic super-cluster vanishes in 10^{117} years. Beyond this inconceivable time-horizon, the cosmos contains matter-particles only in the form of positronium, neutrinos, and gamma-ray photons.

Just when matter exits in the universe depends on whether or not protons decay. If they do, protons and the products left over from the decay of other baryons (heavy "matter"-particles) vanish in supergalactic black holes at the time-horizon of 10^{117} years. If protons do *not* decay, this horizon expands to 10^{122} years. At that time even nondecaying protons evaporate in the last of the remaining giant black holes.

The fate of matter in the universe will also seal the fate of life. Indeed, the complex configurations required for the phenomena of life will vanish much before matter itself decays.

In a *closed* universe—one that ultimately collapses back on itself—background radiation will increase gradually but inexorably. The wavelength of radiation will contract from the microwave region into the region of radio waves, and then into the infrared spectrum. When it reaches the visible spectrum all of space will be lit with an intense light. All life-bearing planets will then be vaporized, along with other celestial bodies.

In an *open* universe—one that expands indefinitely—life will disappear because of cold rather than heat. As galaxies continue to move outward, many active stars will complete their natural life cycles before gravitational forces bunch them close enough to create a serious risk of collision. But this does not improve the long-term prospects of life. Sooner or later all active stars will exhaust their nuclear fuel, and then their energy output must diminish. The dying stars will either expand to the red-giant stage, swallowing up their inner planets, or settle into lower luminosity levels on the way to becoming white dwarfs or neutron stars. At these lower energy levels they will not be capable of sustaining whatever life has evolved on their planets.

Big Bang cosmology differs from most of the historical conceptions in that it maintains that whatever came before the Big Bang, and whatever will follow after the Big Crunch (or after the evaporation of the last galactic-cluster-size black holes), is unknown, and it is intrinsically unknowable. Do not ask about it: the question, cosmologists say, is meaningless.

But Big Bang cosmology may not be the last word. The cosmos may not have been born some 15 billion years ago, and may not end even at the stupendous time-horizon of 10^{117} or 10^{122} years. There may have been cosmos before the Big Bang, and there may be cosmos after the last particles of matter that were synthesized in its heat have decayed. Indeed, evidence is surfacing that there are particles and entire galaxies in the cosmos that are not the product of the Big Bang, but—perhaps—of prior "bangs" up to hundreds of billions of years ago.

IN SUMMARY . . .

Current cosmology can answer many a perennial question regarding the nature and destiny of the natural universe, but cannot answer them all. Its limitations may be arbitrary, however: Big Bang theory itself may not be the final answer. Not only are there important questions this theory cannot field; there are also a number of observational problems it is unable to resolve. We shall review these problems, and the alternative solutions now offered to them, in Part Two, when we consider the wide gamut of puzzles that still dot the landscape of the contemporary sciences.

CHAPTER 2

THE NATURE OF MATTER

WHAT IS MATTER? At first sight, the question seems naïve and bordering on the ridiculous. After all, our bodies are built of matter, and when we hit our fist on the table we encounter matter in no uncertain form. Common sense tells us that everything that is "really real" is material—the rest, except for the space in which material things disport themselves, is but imagination or illusion.

The matter of matter is not quite as simple as that. Many an ancient mystic and metaphysician held that matter is a complex form of energy, and some ancient notions spoke of it as a condensation of space. In the modern age of materialistic–mechanistic science, such notions appeared as idle speculation or mere superstition. But in the 20th century, scientists have not been as categorical. Despite remarkable insights into the origins and evolution of matter in the cosmos, contemporary physical science has no dogmatic stance on the definition of matter. In contemplating the question, scientists face a mystery that is as deep as the ultimate origins and the destiny of the cosmos, and it is considerably more surprising. After all, the cosmos is very big and very old, and may even be infinite and eternal. Knowing it is not an easy matter. But matter is right here: it is all around us and, indeed, in us. Why can we not know it with more certainty?

The question of certain knowledge is one that philosophers have not ceased to argue ever since the Greeks raised it 2,500 years ago. Without rehearsing the arguments in detail, let it be said that almost all philosophers, and practically all scientists, would now subscribe to the thesis that nothing that we observe in the world, including our own bodies, can be known with absolute certainty. Doubt as to what it is, and even *whether* it is, can never be fully excluded. Only the fact of our own mental processes, including the very act of doubting, is indubitably certain. As Descartes said, *cogito ergo sum*—I think therefore I am. *That* I think, at least, is certain. But that the "I" to which this statement refers would be a material entity neither Descartes nor any other philosopher or scientist could assert with complete assurance. The mere fact that I can hit my knuckles on the table and feel the pain, and that I can pinch my cheek and feel the pressure, does not necessarily mean that I come across matter. My body could still be a densification of some subtle substance such as energy, and if it encountered another densified energy in the shape of a table it would not be able to penetrate it. Less dense energies (such as water) could be penetrated, though not without resistance, while more rarefied energies (such as air) would offer less resistance—that is, *if* matter is truly energy, which is by no means to be assumed without further evidence.

But the proposition that matter in its ultimate nature is not "material" but something else, is intrinsically reasonable. The world need not be built of tiny hard and indivisible building blocks, like minuscule bricks or billiard balls. It could also be built of energy fields, or something different still. The entire issue calls for a deeper look.

In the history of science and philosophy, in the East no less than in the West, the question was subjected to many a deep look. In order to frame the concept of matter that emerges in the new physics, we should review the main thrust of these historical arguments. They raise the alternative possibilities for the way we human observers can approach and understand the question concerning the ultimate nature of physical reality.

THE PERENNIAL SEARCH

The search for understanding matter has always been bound up with the search for understanding the basic nature of reality. The search began in the great civilizations of antiquity. In the 6th century BC the Ionian philosophers of nature divested themselves of the mythological worldviews that had dominated Mediterranean civilization until then, and attempted to comprehend the nature of the world in terms of its origins from a shared fundamental stuff or substance. In the framework of their cosmological speculations (already noted in Chapter One), the early natural philosophers did not draw a radical distinction between matter and mind, and material and ideal reality; they reasoned that all the diversity and order that now meets the eye must have arisen in the course of time from a state of lesser diversity and more disorder. This process, they thought, has a logic and unity all its own.

The first attempts were centered on understanding the variegated world of sense experience in terms of an underlying unity called "the One." The One was to be found in a grain of sand as well as in the totality of the universe. The microcosm reflects the macrocosm; the macrocosm shines forth in the microcosm. The Greeks were also aware of "the Many": they saw the great variety of things in the world, the plants, the animals, the people, as well as the sea and the clouds. They explained this diversity as emerging from a basic original "stuff" or "substance": unity, they said, is always present in the womb of diversity.

According to Thales, the original unitary substance was water, while his disciple Anaximander suggested that fire, earth and air played an equally important role: the original substance was undefined, limitless, and all-encompassing. Anaximenes, in turn, maintained that the primeval substance was a mixture of water and earth that, warmed by the Sun, generated plants, animals and human beings by spontaneous creation.

The rational minds of the Greeks developed the kind of natural philosophy initiated by Thales to a high degree of sophistication. Heraclites, who thought fire the most important of the substances, placed stress on eternal becoming: on the principle that "change is all." In his famous words, one cannot step into the same river twice; one cannot know any one thing in the world for what it truly is— it is constantly changing. Empedocles, in turn, viewed all things as composed of air, earth, fire and water in measures determined by the principle of love, which binds, and of hate, which separates. From fires within the Earth arose the primal forms that later evolved into the familiar organisms. Many were imperfect and disappeared, while those that proved perfect survived.

With Socrates the naturalism of the Ionian philosophers became tuned to the human world: man is the measure. Plato, his great disciple, maintained a view of reality only as a "likely tale": he considered things in the perceived world to be images or shadows of eternal and unchanging Forms or Ideas. Aristotle replaced this conception with a naturalism in which careful observation is integrated with a truly encyclopedic range of knowledge. The Aristotelian "great chain of being" extends from inanimate objects through plants and animals all the way to humans. In it the progressive development of nature is matched by a ripening of the soul. The inorganic becomes the organic through metamorphosis, and in the realm of the organic, animals endowed with powers of sensibility are more animate than plants, which are endowed merely with the powers of nourishment. Nature proceeds gradually and constantly from the least to the most perfect, becoming ever more complex in the process. The progression is not accidental: nothing, according to Aristotle, exists without a cause. Nature's process is pulled as it were by a final cause, which is perfection.

It was Leucippus and Democritus who advanced the theory of

matter that was to make the deepest impression on modern science. All things, said their theory, are made up of atoms: indivisible
and indestructible building blocks of the real world. Atoms and all
things composed of atoms constitute the sphere of Being, but since
atoms can and do change, Being cannot fill all of space—there
must also be a Void: the sphere of non-Being. Change can occur in
the world because atoms in the course of their existence adopt
different positions and form different things in the Void.

In the atomist theory the Democriteans grasped what many
believed to be the real nature of the world—that it is made up
of elementary and indestructible elements of matter that in combination form all there is. In one version or another, this theory
endured for almost 2,000 years; it fell apart only with the rise of
experimental physics in the 19th century.

Science in the experimental mode arose at the dawn of the
modern age, when the Renaissance and the Reformation weakened
the hold of Christian doctrine on the European mind. Independent
inquiry began outside the monastic walls, despite the objections of
the Pope and the persecution of Giordano Bruno and Galileo.
Hesitantly but inexorably, scientific research took off: a civilization
based on the theories and practical applications of experimental
science came of age.

Given the relatively primitive means and instruments available
to the early experimental scientists, the first theories were based on
problems that could be solved with the help of the methods and
the equipment they commanded. These were theories that postulated universal laws of motion inferred from the observation of the
speed and trajectory of falling bodies and the acceleration of balls
as they rolled down inclined planes. Not surprisingly, the worldview of modern science became the view of a giant mechanism
that, despite its complexity, obeyed simple and basic laws.

The mechanistic view, first articulated by Galileo, contrasted
sharply with the world of the living, infused as the latter seemed to
be with purpose and consciousness. The result was the inexorable

divorce of natural science from human and spiritual concerns. At the dawn of the modern age, the Western systems of knowledge split into natural philosophy (which was then empirical science as a whole) and moral philosophy (which embraced the speculative extensions that came to be known as the humanities).

Natural philosophy culminated in Newton's magnificent synthesis. His *Philosophiae Naturalis Principia Mathematica*, published in 1687, demonstrated with geometrical certainty that material bodies on Earth move according to mathematically expressible rules, while the planets in the sky rotate in accordance with Kepler's laws. The motion of all things was shown to be rigorously determined by the conditions under which it was initiated, just as the motion of a pendulum is determined by its length and its initial displacement, and that of a projectile by its launch angle and acceleration. With mathematical precision Newton's classical mechanics predicted the position of the planets, the motion of pendulums, the path of projectiles, and the motions of all the "mass-points" that replaced Democritus' atoms as the ultimate building blocks of the scientifically known world.

Newton himself, however, did not believe that geometrical descriptions and mathematically formulated laws regarding matter and force could give a complete description of reality. A hermetic magus and religious mystic, he suspected that God plays an active role in nature, and was convinced that these divine interventions could only be grasped through mystical insights and esoteric calculations. But Newton did not have time to include his mystical and numerological arguments in the *Principia*—he intended to add these subsequently in a detailed discussion of the pertinent religious and mystical doctrines. However, as the revised edition neared completion—it was held up by its author's indisposition following the original publication (he appears to have suffered a nervous breakdown and did not work for nearly two years)—Newton was confronted with the realization that he was famous throughout the Western world precisely *because* he explained

physical reality without reference to God, alchemy, or other esoteric means. The dilemma, whether to let the treatise stand as it was
and merit Alexander Pope's homage ("God said, 'Let Newton be'/
And all was light") or to "complete it" with his religious views and
mystical doctrines, could not have been an easy one. In the end,
the *Principia* remained unchanged.

Newton's fame rested on understandable foundations. In addition to marvelous elegance and apparent completeness, his
classical mechanics provided just the kind of certainty European
civilization needed after the shock of the Black Death and the un-
easy doubts it triggered regarding God's goodness and omnipotence.

In the centuries that followed, Newton's synthesis became an
all-embracing paradigm for thought and action, unfailing and unchallenged. Rationally-thinking people felt confident that they
could at last unlock the mysteries of nature. As Fontanelle put it at
the end of the 17th century, the world is like the face of a clock.
For ages, people had marveled at the movement of the clock's
hands and made up stories about what could drive them with such
precision and regularity. Now, thanks to Newton's physics, they could
look behind the face and see the gears and the pulleys at work.

The simile was well taken. The world of Newtonian physics
was a clockwork universe, rigorous and determined, and forever
unchanging in its obedience to the basic laws of motion. The
knowledge conveyed by it seemed practically total. The mathematician Laplace boasted that an intelligence with a knowledge of all
the forces controlling nature, together with the momentary state of
all nature's entities, "would be able to embrace in a single formula
the movements of the largest bodies in the universe and those of
the lightest atom; for it, nothing would be uncertain; the future and
the past would be equally present to its eyes."

THE CONTEMPORARY COMPLICATIONS

Though unchallenged since the end of the 17th century, Newton's
grand synthesis began to collapse at the dawn of the 20th century.

We can trace the problems back to the time when, in the beginning of the 19th century, the atomist theory of Democritus was rediscovered by the English chemist John Dalton. Dalton's theory, that all gases are made up of small and indivisible units called atoms, created a revolution in chemistry. Its triumph, however, was short-lived. Within only 50 years of the publication of Dalton's theory, experimenters discovered that atoms are not indivisible but are made up of still smaller particles. Even these particles could not be the ultimate "atoms" the Greeks spoke about, because if they have finite extension in space, they must be further divisible. Indeed, when sufficiently powerful experimental devices became available, not only atoms but even their nuclei proved fissionable.

With the splitting of the atom in the late 19th century, and of the atomic nucleus in the early 20th, more had been fragmented than a physical entity. The entire edifice of classical natural science was shaken. The experiments of early 20th-century physics demolished the view that all of reality is built of indivisible atoms, but physicists could not put any comparably coherent and meaningful concept in its place: the very notion of matter had become problematic. The subatomic particles that emerged when atoms and atomic nuclei were fissioned did not behave like conventional solids: they had a mysterious interconnection known as "nonlocality," and a dual nature consisting of wave-like as well as corpuscle-like properties.

By the 1920s, scientists in the emerging field of quantum physics faced a world in which physical reality became strange beyond all expectations. Space and time, rather than passive backdrops for the intercourse of material atoms (or Newtonian mass-points), became complex entities in their own right, interacting with photons and electrons, and entering into the very texture of physical phenomena. To philosophers and philosophically-minded scientists it appeared that the physical universe itself had dematerialized: it had become, in philosopher Sir Karl Popper's words, more like a cloud than a rock.

The revolution that came with quantum physics in the 1920s was even more radical than that which occurred with relativity physics at the turn of the century. Einstein's physics preserved the unambiguous description and fundamental determinism that had characterized Newtonian physics. Quantum theory, on the other hand, did away with unambiguous paths of motion and introduced probabilistic indeterminacy into the very foundations of material reality. From then on, the domain of matter became more and more mysterious. Objective reality appeared to dissolve before the eyes of the wondering quantum physicists. Faced with the enigmas of nature, many scientists, led by Danish physicist Niels Bohr, decided to suspend speculation concerning the independent nature of what they were observing: they considered the objects of their observations merely as "phenomena."

Phenomena, German physicist Werner Heisenberg noted, are not the "works" of nature but merely the "texts" of science. "The atomic physicist," said Heisenberg, "has to resign himself to the fact that his science is but a link in the infinite chain of man's argument with nature, and that it cannot simply speak of nature 'in itself'." "We are suspended in language," Bohr concurred, "physics concerns what we can say about nature." The external world of physics, according to Arthur Eddington, had become a world of shadows. "Nothing is real," he wrote, "not even one's wife. Quantum physics leads the scientist to the belief that his wife is a rather elaborate differential equation." (But, Eddington added, he is probably tactful enough not to obtrude this opinion in domestic life.)

Though advised not to think about the nature of reality beyond the scope of their observations in the laboratory, some physicists ventured further. They speculated that the world to which language and the "text" of science refers is mental rather than material. "To put the conclusion crudely," said Eddington, "the stuff of the world is mind-stuff." Jeans agreed. ". . . the cumulative evidence of various pieces of probable reasoning makes it seem more and more likely that reality is better described as mental than as material . . .

the universe seems to be nearer to a great thought than a great machine."

Heisenberg spoke regretfully of the error of the "philosophical doctrine of Democritus." The world, he said, is built as a mathematical, not a material, structure; there is no use asking to what, beyond themselves, the formulas of mathematical physics would refer. Much as Plato had dissolved the materialism of the Ionian natural philosophers in the abstract world of forms and ideas, so the deterministic world of classical mechanics was now to be dissolved in the complex formulas of mathematical physics.

Not only could scientists not identify the basic entities that would underlie the diversity of manifest phenomena, they could not even say whether any such entities existed in nature. Clearly, neither the Democritean atom nor the Newtonian mass-point was the ultimate ground of physical reality. In Hungarian-born Princeton physicist Eugene Wigner's apt phrase, modern quantum physics had to content itself with dealing with "observations" rather than "observables." Physicists could describe what they observed, but could not refer them to realities that would subsist independently of their being observed. The situation resembled that which Alice found in Wonderland: matter-particles, like the Cheshire Cat, exhibited a grin, but there was nothing that could carry their grin.

This state of affairs was not a happy one, and was by no means accepted by everyone. The puzzles encountered by quantum physicists in the laboratory inspired the most famous and prolonged in-depth discussion on the nature of physical reality in the history of modern science. In the years from 1927 to 1933, Albert Einstein and Niels Bohr met together periodically and corresponded on the interpretation of the puzzling observations. Einstein could not abide by the strange indeterminacy that seemed inherent in the behavior of the elementary particles; he brought up one thought experiment after another to show that quantum theory as it was then formulated was logically inconsistent. Bohr, in turn, refused any interpretation that went beyond the range of actual observations. Nature, Bohr claimed, has placed an absolute limit not only on what can be measured and observed, but also on what one can speak about without ambiguity.

Einstein agreed to the Heisenberg indeterminacy principle—that both the position and the momentum of an elementary particle cannot be measured simultaneously—but did not concede that this would mean that elementary particles did not have a definite position and momentum at all times. Bohr disagreed; in his view it did not make sense to speak of a particle having a definite trajectory in the absence of its "registration" by an observer or an instrument. Without being registered by an observer (or perhaps just by an instrument), the particle, said Bohr, could not be said to exist on its own. This view, however, was not acceptable to Einstein. "If a person, such as a mouse, looks at the world, does that change the state of the world?" he asked in physicist John Wheeler's relativity seminar at Princeton. "I find the idea quite intolerable," Einstein wrote earlier to Max Born. "If the existing interpretation would prove correct," he continued, "I would rather be a cobbler, or even an employee in a gaming house, than a physicist."

In the final phase of his dialogue with Einstein, Bohr found himself restricted to the term "quantum phenomenon." This term, Wheeler pointed out later, is highly significant. It suggests that in speaking of a particle we are no longer dealing with an objective, observer-independent reality. We have no basis for speaking of what particles *are* and of what they are *doing* between the observations that signal their emission and their reception. Whatever occurs in between is, in Wheeler's picturesque phrase, a "great smoky dragon." Its tail is sharp where the particle is emitted, and its mouth is sharp where it bites the detector, but its body in between is "smoky." The "quantum phenomenon," said Wheeler, "is the strangest thing in this strange world."

For the most part, present-day quantum physicists accept this strangest of all phenomena—their equations, the pride of three generations of uncommonsensical investigators, do the work expected of them. As a rule, physicists do not venture to search for more realism if the search involves questioning the validity of the mainstream formulas. The fact is that, after some 70 years of research and exploration, quantum theory can take credit for being both stupendously successful and excruciatingly puzzling. In the exploration of the subatomic world it has been used by thousands of physicists in almost every conceivable experiment, and it has performed remarkably consistently. At the same time, it has left large gaps in our understanding of what any healthy mind could accept as the nature of an observer-independent reality.

For one thing, quantum theory cannot describe the basic constituents of familiar objects as elements within those objects. It speaks of commonsensically inconceivable, probabilistic entities that are present simultaneously in more than one location, and are either waves or corpuscles, according to what questions we ask of them—and how we interact with them. For another thing, quantum theory also fails to account for one of our most basic intuitions about ourselves and the world at large: the irreversible passage of time.

In the context of the quantum formalisms, the irreversibility of time is a ghostly chimera—a consequence, in Hungarian-born mathematician John von Neumann's view, of our acts of measurement. The past and the future in the subatomic domain can no more be distinguished by Erwin Schrödinger's equations of the quantum state of the particle than the past and the future of macroscopic bodies and processes can be distinguished by Newton's equations of motion. In the mathematics of quantum theory, time enters only when the intrinsically probabilistic and well-nigh inconceivable state of the particle resolves into a definite state familiar from our everyday commerce with reality—that is, when the particle's uncommonsensically superposed wave-function "collapses" into a commonsensically deterministic state. That collapse, however, and with it the resolution of the superposed probabilities, is not a feature of reality independent of our observation: it must be attributed to our *interaction* with reality—in von Neumann's view, to our act of measuring the particle, or, in Eugene Wigner's concept, to the interaction between the particle and our conscious mind.

IN SUMMARY . . .

Cosmological and relativity physics have penetrated to the far reaches of space and time, but the ultimate basis of physical reality, as expressed by French physicist Bernard d'Espagnat, remains "veiled." Notwithstanding the indubitable successes chalked up by quantum physics in regard to computing processes that occur in the small-scale structures of the physical world, the entities that would populate that world continue to be fuzzy, described by unvisualizable and realistically inconceivable mathematical formalisms. These may suffice for purposes of computation, but hardly for those of meaning. When all is said and done, quantum theory remains essentially incomplete. As of today the mainstream theorists of the quantum world have not succeeded in giving an unambiguous answer to the question, "What is matter?"

CHAPTER 3

THE PHENOMENA OF LIFE

L IFE IS AT ONCE the most familiar and the most mysterious of natural phenomena. As long as we think and breathe, we cannot doubt that we are alive, but this certainty does not include an answer to the other fundamental query "What is life?"

If the nature of matter remains veiled, and if there are unsolved problems in our understanding of the nature of the cosmos, what can we expect of science's understanding of the nature of life? Life, after all, could be the workings of a particular complex chemical machine, assembled from millions and billions of atoms, molecules and cells. It could also be the manifestation of a reality that is entirely different from that of the physical world—a reality that is spiritual in its essence. The Western religions speak of an immortal soul which is only temporarily associated with the living body and is destined for eternal life—or ceaseless damnation. And the Eastern religions hold that living bodies are infused time and time again by a nonmaterial element such as a soul, with its acquired merits and demerits carried over from life to life as karma.

For many people, especially in the West, science means rejecting such notions. Western common sense is more comfortable with the view that the living organism is a complex machine. But the new sciences of life have gone way beyond this concept. Though scientists (with occasional exceptions) do not maintain the notion

of a soul or other "life principle" associated with an otherwise lifeless body, they have an entirely different notion of what we are
to understand by a living organism. We can best understand this
notion by placing it into the historical perspective—taking a look at
how it actually came about.

LIFE FROM NONLIFE

To trace the evolution of the modern concept of life, we need not
go back to Adam and Eve; it is enough to return to the mid-19th
century. Here we find classical mechanics, the dominant natural
science of the day, in difficulty. Its laws of motion apply to mass-
points moving in space and time, and the space in which they
move is Euclidean (three-dimensional and flat) and the time in
which they move is reversible—in Newtonian physics every reaction that can proceed forward can also proceed backward. As we
have seen, Newton's laws explain the motion of the pendulum, the
fall of objects to the ground, and even the movement of the planets
around the Sun. But they do not explain the phenomena of life.
Living systems manifest time-irreversible processes (otherwise
they would be eternal), and their "motions" are too complex to be
computed with the tools developed for the measurement and computation of physical systems. Also their origins are wrapped in
mystery. If Darwin is right and they evolved out of the physical
world, that world must be infused with some extra-physical principle, something like Henri Bergson's *élan vital*.

The puzzlement of 19th-century scientists shows how deeply
the mechanistic universe of Newtonian physics was at odds with
Darwin's evolving world of life. The physical universe was known
to be a time-reversible mechanism, governed by a small number of
universal laws, while the living world was believed to evolve in an
irreversible fashion. Science's concept of reality was torn asunder.

There was, however, a discipline that, unlike classical mechanics,
affirmed rather than denied time's irreversibility. This was classical
thermodynamics, of which the justly famous second law states that

in a closed system that performs work, disorder and randomness can only increase. (Closed systems do not interchange matter and energy with their environment, hence whatever free energy they start with, must be ultimately used up.) This tenet, however brilliant and exact its formulation, has been of little help in bridging the rift between Newtonian physics and Darwinian biology. Time's arrow, though it now flew in the universe, flew in the wrong direction. The observed world should be running down, instead of winding up. But life seemed to do the latter: it began with the lowly cell and simple algae and sponges that are built from associations of cells, and continually climbed the ladder of complexity until it reached the animal kingdom, topped, it seems, by man.

In the course of life's evolution, organisms became more complex, rather than less. And the second law of thermodynamics did not explain how this would have come about—even if its formulas did not actually forbid it.

As we look back on it today, we find that 19th-century natural science was saddled with two arrows of time (the thermodynamic and the biological)—and an undisputed physical framework (the Newtonian) that did not explain either. The scientific community had to wait until the second half of the 20th century for an acceptable resolution of the contradiction. By the late 1950s the new discipline of "nonequilibrium (as contrasted with classical) thermodynamics" showed that living organisms are not the kind of systems that must inevitably run down: they are not closed systems. Living systems are essentially *open* systems, and because they constantly replenish their store of free energy, they can maintain themselves far from the inert regime of thermodynamic equilibrium—like an engine that can keep running when it is fed with fresh fuel.

In light of these concepts, life, though far from being "nothing but" a fanciful variety of physical system, can be seen as a logical continuation of the processes that occur in the physical universe. The evolution of the universe brought about galaxies, and within the galaxies stars, some with planets. Some planets happened to be at an orbit where the energy flow from the mother sun permitted the further structuring of the already complex chemical soup on their surface. Constant irradiation with energy from the local sun energized the chemical soup, and created open systems—systems with a flow of energy through their boundaries—and these systems moved ever further from the inert state of chemical and thermal equilibrium. On suitable planetary surfaces, like those of Earth, life evolved in a condition far from equilibrium without benefit of extraphysical principles such as life forces and immaterial souls.

According to the standard BB (Big Bang) theory, the universe must have been at least 10 billion years old before life appeared on this planet (and perhaps others). It appears that the protoplanetary gas around the Sun began to solidify about 4.56 billion years ago and, remarkably, the evolution of life on Earth seems to have begun not long after. Fossils with traces of advanced chemical evolution dating back 3.5 billion years have been identified; primitive biological organisms have been shown to have existed at least 2.8 billion years ago, and evidence of the biochemical activity of prokaryotic (nonnucleated) cells with modern enzyme structures has been uncovered in fossils 2.3 billion years old.

The chemical constituents required for the evolution of life were present in Earth's chemical soup before biological evolution took off. The six elements that make up some 98 percent of the known universe—hydrogen, helium, carbon, nitrogen, oxygen and neon—together with the more complex molecules essential for the synthesis of the first self-replicating cells, were already synthesized in the cosmos. Even amino acids and nucleic acids have been (and probably still are) created in the universe; they have been found in meteorites. For this reason it appears more than likely that life will have evolved in some form on other planets as well: there are billions of them in our galaxy, and there are billions of other galaxies.

Thermal and chemical conditions on our own planet were highly suitable for the synthesis of the more complex molecular constituents of life. Monomers like sugars, amino acids, purine and pyrimidine bases, and linear polymers built of these monomers such as proteins, nucleic acids and other macromolecules, could be synthesized in the constant energy flow from the Sun. In time, the prokaryotic cells that preceded the higher forms of life evolved and became integral elements in the emerging biosphere of the planet.

According to current conceptions, the higher forms of life evolved when the globally dominant community of algae (consisting of prokaryotic nonnucleated cells) was destabilized by the appearance of single-celled eukaryotes that fed on the algal community. By cropping the algae they broke through the epoch of stasis that had persisted a billion years or more. The algae were destabilized; niches for additional species were created; and subspecies that emerged on the periphery were able to move in and occupy them. A large variety of prokaryotes appeared, and these in turn made possible the emergence of more specialized eukaryotes which functioned as their "predators."

With the sole exception of viruses and bacteria, the organic species that now populate the Earth have descended from the early eukaryotic cells. Entire families of species—so-called genera—emerged in the bursts of creativity that marked successive evolutionary epochs. Some 600 million years ago the Cambrian explosion brought into being most of the invertebrate species within the relatively brief span of a few million years. *Homo* himself emerged in this vast evolutionary development, though he was a latecomer.

In the generally irreversible evolution of species, those that were capable of surviving under a wide range of environmental conditions (such as a variety of climates, topologies, predator and prey populations, etc.) survived longer than those that had fitted themselves into narrow environmental "ruts." Specialist species mutate and die out under many conditions where the generalists can adapt and survive. As a result the diagram of the tree of life no longer resembles the continuous Y-shaped joints of the classical Darwinian theory; it is now pictured in terms of abrupt switches, from dominant species which become extinct to hitherto peripheral ones that become dominant. The specialists have short life lines (they keep dying out, to be replaced by mutants), while the generalists have relatively long ones.

Since its origins in shallow and warm primeval seas, organic life has been sustained by the flow of free energy from the Sun. Plants use sunlight in photosynthesis, converting water and carbon dioxide into carbohydrates; animals eat plants or other animals, and humans at the top of the food chain eat both plants and animals. Were the energy differential between the surface of the Sun (approximately 6,000 degrees C) and the surface of the Earth (about 25 degrees C) ever to equalize, not only life but all thermodynamic processes on the planet would soon come to an end. The heat stored in the Earth's atmosphere would be depleted in a few months, while that in the oceans would be dissipated in a matter of weeks. Only worms and clams at the bottom of the deepest oceans would survive for any appreciable amount of time. However, as long as the stream of energy from the Sun to the surface of our planet lasts—and it is expected to last for billions of years more—living systems will convert some of the free energy of that stream into biomass. They will not only maintain their structure but will also evolve new and different structures, some more complex and sophisticated than those that existed previously.

THE DRIVING FORCE OF EVOLUTION

The processes of evolution on this planet manifest an unsuspected power: in the course of the last 3.5 billion years they have created a mass of living cells of which the combined weight adds up to more than the weight of the six continents. This enormous mass of organic material not only kept reproducing itself, but became ever more complex. The rhythms of evolution have been speeding up throughout the tenure of life on Earth. Biologist William Day noted that over half of evolutionary time was taken up with the advance from one stage to another of cellular life: from the nonnucleated prokaryotes to the nucleated eukaryotes. It took evolution half that time again to reach the level of fish. Then, as the succeeding steps followed, time intervals between the major innovations shortened. While sooner or later some segments of the living world achieved a form of balance with their environment and ceased evolving, the overall advance of the evolutionary wave did not cease: it continued to accelerate stage after stage.

The progressive steps and stages of evolution have to be described in terms of shorter and shorter periods marking the major time intervals. (The principal stages in the evolutionary process are denominated *eons, eras, systems, periods, epochs,* and *ages,* with eons denoting the longest time periods and ages the shortest.) Events at the beginning of biological evolution are classified under a single heading: the Azoic ("without life") era. During this interval the crust of the Earth melted and hardened repeatedly, ultimately forming a permanent crust enveloped by the atmosphere and covered for the most part by the oceans. Then, in the next, so-called Archeozoic ("primeval life") era, bacteria and algae, the first and most primitive forms of life, appeared. The Proterozoic era, marked by widespread glaciations, floods, and a great movement of land masses, saw the emergence of simple invertebrate animals. In the Paleozoic era, fish and reptiles, as well as the first forests, made their appearance, while the Mesozoic era included the rise and fall of the dinosaurs and the evolution of birds and modern plants.

The Cenozoic is the most recent era, and the evolutionary burst it created was so diverse that it had to be classified under separate "systems." These are the Paleogene, comprising the evolution of mammals and other modern varieties of animals; the Neogene (further subdivided into the Miocene and the Pliocene), where living forms continued to specialize and diverge, and the Quaternary, which is the most recent of the major evolutionary periods. This later period is divided in turn into the three familiar ages: the Lower, the Middle, and the Upper Pleistocene.

Throughout the period that life has evolved, time has shrunk dramatically. Evolution seems to describe the curve of an accelerating object building momentum. The driving force goes unchecked; momentum sets the pace—like a stone dropped from a height. The Archeozoic lasted from between 3.5 and 4.5 billion to about 2 billion years before our time; the Proterozoic began 2–2.5 billion years ago, and gave way to the Paleozoic little more than half a billion (550–600 million) years in the past. The Mesozoic is dated from about 200–250 million years, and the first of the Cenozoic Systems began some 65–70 million years ago. Then evolution accelerated to yet another dimension. The Miocene epoch is about 25 million years in the past, the Lower Pleistocene of the Quaternary began 1.6 million years, the Middle 750,000 years, and the Upper but 125,000 years before our time. Hominid creatures appeared during the Holocene (or Recent) epoch, though our lineage may have diverged from other hominoid species much before then.

THE ENTRY OF *HOMO*

Hominoid species embrace three families: the hylobatides, of which the living representative is the gibbon; the pongids or great apes, currently represented by the orangutan, the chimpanzee and the gorilla; and the *hominids*. The latter is the family of man (though in some conceptions the chimpanzee and the gorilla are seen as belonging to this family, under the assumption that our forebears branched from them only more recently).

Homo separated from the other two hominoid families when the early hominids descended from the trees. Why they did so is still a matter of conjecture—the move may have been prompted by changes in the climate. Due to major shifts in the continental plates 5 million or more years ago, there were likely to be major alterations in air movements and consequently in weather patterns. In south and central Africa the tropical forests retreated and lush vegetation became more scarce. The early bands of hominids may have found themselves pushed more and more to the ground, looking for fruit, low shrub, and root foods. Colonies near forest margins would have been obliged to travel increasing distances between wooded areas. Under these conditions those who could move on two legs had a distinct edge on survival.

Another factor favoring bipedal locomotion may have been the safety of infants. Among tree-living primates a major cause of infant mortality must have been newborns falling to the ground because they failed to cling to their mothers. Tribes in which females could hold their infants with their forelimbs would have had a reproductive edge: more of their young would have survived. As such females would have been less agile in the high branches, they would have preferred to spend more of their time on the ground. There, holding an infant with one forelimb and picking roots and berries with the other, those who could walk on their hind legs had the advantage.

Although factors such as the above no doubt played a major role in the switch from tree-dwelling apes to an upright species on the ground, a host of further factors would have conspired to carry through this transformation. On the ground an erect posture was needed to provide early warning of danger, and a flat-soled foot offered a margin of safety when escaping from stronger and faster carnivores. Longer and straighter leg bones, with a large toe that could take much of the body weight and push off each step, carried an obvious advantage. Forelimbs, freed from having to hold on to branches in the trees, could then be used for other functions. Arms and finger bones became straightened, and thumbs extended and counterposed to the other fingers. These modifications allowed a firm grasp and the increasingly precise manipulation of a variety of objects. At the same time the size of jaws was reduced: they were no longer needed for grasping and food gathering. The entire skeletal structure became less massive, including the bone mass that surrounded the skull. A larger cranium could—and did—develop, housing a larger brain.

About 1.6 million years before our time an upright large-brain species appeared, appropriately named *Homo erectus*. It had the ability to make hand-axes and to use fire. During the next 600,000 years it spread from Africa throughout Asia and Europe. *Sapiens* was one of the descendants of *erectus,* appearing in the fossil record some 50,000–100,000 years ago. Another branch, *sapiens Neanderthalensis,* or Neanderthal man, appeared at the same time but has left no trace of its existence for the past 35,000 years. Thereafter the modern form of *sapiens* (*Homo sapiens sapiens*) has been the sole representative of the hominid lineage on this planet.

IN SUMMARY . . .

While it would be an exaggeration to claim that science has resolved the question concerning the nature of life, its conception is a vast improvement on prior speculations. Life is no longer a stranger in the universe, but an emergent part of it. A living organism is not "just" a physical system, but it is also not a completely nonphysical one. Life is a product of the long-term development that structured the cosmic fireball into hadrons and leptons, and stars and galaxies—and that, in our local star-system, further structured the rich chemical soup on Earth's primeval seas into self-maintaining, open, thermodynamic systems. The living systems that emerged could feed on the free energy they obtained in the constant flow of radiant energy from the Sun, cycling it through a complex chain that relates the lowliest algae to the highest predator. After 3 billion years of accelerating evolution, the system of life on this planet worked itself into a seamless totality with capacities for self-regulation that come close to homeostatic processes in living organisms.

English biologist James Lovelock maintains that Gaia, the system of the biosphere and its physical environment, is a living organism in its own right. Be that as it may, it is clear that the web of life on Earth is a highly harmonized whole. *Homo* has entered into, and is now living in, this finely tuned system, whether he realizes it or not.

CHAPTER 4

THE MANIFESTATIONS OF MIND

T HE LIVED, immediately-perceived stream of conscious experience that accompanies each of us throughout our life is an enduring object of poetical wonderment as well as philosophical debate. In contemplating it, we have reached the outermost bounds of scientifically researchable questions. Though mind is the most immediately and intimately known aspect of our experience—indeed, in some views it is the sum total of our experience —the further fundamental question "What is mind?" is not easily answered. Our consciousness seems to float in our head; no amount of introspection can reveal the gray matter with which natural science believes it to be associated.

The notable fact that has always struck philosophers and philosophically inclined scientists is that such a stream of consciousness should be associated at all with a brain and body that consists of tissues, organs, bones, and gray matter. These bodily parts are made up of cells, which in turn are made up of molecules and atoms; and the molecules and atoms of the brain do not show any evidence of being conscious, or different in any way from other molecules and atoms. Yet consciousness somehow "infuses" the atoms and molecules that constitute the physiological gray matter in the brain. Is it that the neurons in our brain are capable of *generating* consciousness? Or is consciousness and the experience of mind an indication of the existence of something entirely different

—a spirit or soul that is not the brain itself but only associated with it?

One view, often maintained by scientists, is that mind is identical with brain: the phenomenon of consciousness is in some way generated in the complex interactions of myriad highly organized brain cells. If so, the question is how physiological gray matter can be so organized that consciousness is effectively generated by it. If such organization is possible, we must admit in principle that artificial systems such as computers built not of neurons but of electronic switches, would likewise generate some form of consciousness. After all, the electronic switches of which they are composed operate in the binary mode—they are either "on" or "off"—much like the neurons of the brain that either "fire" or do not fire.

The contrary view, maintained by mystics, poets, and speculative philosophers, is that there is something specific in the human brain that makes it uniquely capable of generating consciousness. This argument is beyond the domain of natural science, the same as the still stronger thesis that consciousness is not reducible to the brain: it is a mental or spiritual principle distinct from, even if associated with, the gray matter in our cortex. Though such views may contain seeds of truth, the natural sciences have nothing to say of them. Their observational and experimental methods could only confirm the existence of a separate consciousness—a spirit, or soul—if it produced observable effects on the brain. These effects would have to be such that the brain itself could not have produced them. But scientists cannot be certain that any given effect observed in the brain could have been produced by something that is not the brain itself. Such certainty could only be had if we knew everything, or well-nigh everything, about how the brain functions. This knowledge is not available today, and is not likely to be available tomorrow, or at any time in the foreseeable future.

Consequently the operative assumption of scientific inquiry into the phenomenon of mind is that it is associated in some way with neural function in the brain (although this assumption is not crucial in the "soft" branches of psychology, where introspection by the subject is the primary source of information). The brain, in turn, is clearly an organ of the body, so that whatever consciousness infuses the brain is *in* the body, and can interact with it in that location.

THE EVOLUTIONARY PATH TOWARD A CONSCIOUS MIND

In the naturalistic perspective espoused by most scientists we can inquire into the reasons why a conscious mind should have been generated in *Homo*. Has such a mind conferred some survival benefit on our ancestor? If it did, a conscious mind would have been naturally selected, much as fins on fish in the sea and fur on animals in cold climates.

There is indeed evidence that certain mental faculties, such as intelligence, are products of long-term evolutionary developments. While some investigators contest that the presence of intelligence requires consciousness (intelligent information-processing, they say, could also occur in the brain on a pre- or subconscious level), it is clear that *some* varieties of intelligent information-processing benefit from the presence of conscious awareness in the subject— for example, the weighing of alternative behavioral options and strategies. It is in regard to these capabilities that consciousness as a new element in the functions of the nervous system could have been favored by natural selection.

Whether or not it is accompanied by anything like a human type of consciousness, intelligence itself is known to be present also in nonhuman species. Many species have developed some forms of intelligence, and would no doubt have developed it further, had they the need and the opportunity to do so. Whales and dolphins have intelligence, but they live in an aquatic environment that is both more stable and friendly than life on land: sea mammals had no need to evolve their intelligence in the way that land-living mammals did. The latter need an intelligence capable of manipulating the immediate environment, since in terrestrial settings survival calls for complex operations. The availability and retention of water, the on-going procurement of free energy, and the maintenance of a constant temperature are essential functions in the behavioral syndrome dedicated to ensuring the integrity of the complex biochemical reactions on which terrestrial life depends. To ensure these functions in competition with species that are better endowed physically calls for considerable sophistication in the manipulation of a mammal's immediate surroundings. A conscious mind is likely to have proved a valuable aid in performing such manipulations.

It is eminently reasonable to think that our hominid ancestors were in pronounced need of manipulative sophistication. When they left the trees, they had to depend for survival on a high level of bodily control, tactile sensitivity, manual dexterity, and the ability to communicate. These functions called for a complex nervous system topped by a large brain.

The pay-off from more sophisticated information-processing and manipulative capacities became tangible 1.5 million years ago, when some hominid bands mastered the control of fire. They learned to throw dry sticks and foliage into a naturally ignited fire to keep it going, and observed that a stick burning on one end is cool enough on the other to be handled. They learned to ignite fires by rubbing stones together to create a spark, or carried a flaming stick from a naturally ignited fire to a more desirable location.

The mastery of fire gave our dispersed ancestral bands a decisive edge in their fight for survival. Fire inspires fear: flames and embers burn feather, fur, and skin on contact. Since the instinctive reaction of animals is to flee, those who mastered fire could use it for their own defense. Fire is also important in assuring a continuous food supply: meat that quickly rots when raw remains tasty and edible when roasted. By roasting their food, our ancestors no longer had to live from hand to mouth: lean periods between hunts and in poor weather were bridged by communal food stocks.

Having mastered fire, *Homo* had an assured path to dominance. Our forebears no longer had to struggle for survival in constant fear of more powerful species: they could establish habitations, protect them, and stockpile staple foods. Fires were tended in distant locations over long periods of time. There are indications of humanly laid fires at such diverse sites as Zhoukoudien near Beijing, Aragon in the south of France, and Vértesszöllős in Hungary. Near Chesowanja in Kenya archeologists have found baked clay next to hominid bones and man-made stone implements; the 1.5-million-year-old clay shows traces of exposure to heat higher than that which would normally occur in a bush fire. The cave at Zhoukoudien harbored a fire that had been tended on and off for some 230,000 years. It was abandoned only when the roof caved in.

Then, some 8–10,000 years ago, people in the Levant began to exercise a significantly higher level of control over their immediate environment: they learned to domesticate a variety of plants and animals. This enabled them to stay put instead of having to follow their food supply. Several river valleys were settled, including those of the Nile, the Tigris, the Euphrates, the Ganges, and the Huang-Ho. Silt deposited by the rivers acted as a natural fertilizer, while periodically flooding waters functioned as a natural system of irrigation. Nomadic bands turned into settled pasturalists—and the rest, quite literally, is history.

Recorded history is based on the exercise of highly evolved mental capacities within organized societies. Social cooperation calls for sophisticated forms of communication to convey intentions and avoid misunderstandings. It was found that hunting, gathering, defense, and the rearing of infants were tasks that could be performed better with a division of labor than by going it alone. In possession of the rudiments of language, our ancestors gained a significant edge in competition with other species. Social behavior became freed from the rigidity of genetic programming and proved adaptable to a variety of changing circumstances. People could work together in pursuing a growing number of increasingly precise and demanding tasks.

Communication through a symbol-using language is a vast improvement over communication by simple sounds. The ability to make sounds is widespread in the realms of life, but it does not constitute speech. Nonhuman species communicate by warnings for danger, calls for mating, and invitations to join in the hunt and, occasionally, in play. Symbol-based speech, as opposed to signals by sounds, lends even primitive tribes considerable survival advantage, for example, in identifying the location of prey, organizing the hunt, finding a mate, and raising offspring.

In the course of millennia, capacities for manual dexterity and tool use came to be joined in the hominid brain with capacities for language and socialization. The genetically-based sign language of the apes transformed into the system of shared symbols typical of human languages.

There was a further, and probably relatively recent, spin-off from this development. Symbol-based language enabled human beings to identify not only things and events in their environment, but also themselves. This laid the basis for the evolution of the human kind of consciousness: a reflexive consciousness that conveys an awareness of both the world around the human being, and of the human being in the world.

THE MODERN UNDERSTANDING OF MIND

As we have just noted, for contemporary scientists the question "What is mind?" is not meaningful in itself: in a naturalistic perspective, mind is associated with cerebral functions. Cerebral functions, however, pose formidable problems of understanding; problems that have been only partially resolved to date. The functions that set off humans from other animals are concentrated in the neo-cortex, the frontal brain regions that were the last to evolve in our species, and they are immensely complex. They comprise perception, response, and regulation, as well as the analysis and retention of information—cognition as well as *re*-cognition. So far only the simpler, more basic functions have come to be well understood: certain elements of perception, motor response, and organic regulation.

Some facts did come to light, however. It has become clear, for example, that in perception the brain does far more than passively receive information from eyes and ears and other sense organs: it integrates the incoming signals with signals already circulating within the cerebral regions, and adjusts the body's sense organs according to the outcome of that integration. In vision, the radiant energy that reaches the retina does not come organized into ready-made images: the so-called optical array is literally "broad-cast" in the electromagnetic spectrum—it is broadly scattered like radio waves. It takes an instrument or lens of some kind to focus and integrate this array into a coherent pattern. This function is efficiently performed by the retina in conjunction with the visual centers of the brain.

The ear is another organ for the sophisticated analysis of finely shaded signals. The inner ear can amplify mechanical vibrations smaller than the diameter of a hydrogen atom into yes/no re-sponses; the incredibly minute amplitude of 10^{-11} meters produces a sensation. It appears that the basilar membrane is not a passive vibratory system, like a microphone driven by a sound signal; it has additional mechanisms that sufficiently sharpen minute excitation patterns so that they are discriminated. The ear operates in a pas-sive resonance mode only at high signal levels; at low levels it "locks on" to the incoming signal by producing a vibration of its own. This means that the finer ranges of auditory perception in-volve an interaction between the signals coming to the ear from the outside, and the signals produced by the ear itself. Hearing in humans results from the analysis of the phase coherence of the ear's external and internal oscillators.

Analyzing the information that reaches the organism from the outside world (that is, *perception*) is only a small part of human mental faculties: an important part is *cognition,* which involves analysis, and *re-cognition,* which calls for retaining and recalling what has been analyzed. The latter is the function of memory.

In order to store perceptions over time, it would seem that some sort of trace or "engram" must be created in the brain in response to the incoming signals. This would modify the neuronal connections in the analytic networks and produce something like a replayable record. In the words of Nobel laureate neuroscientist Sir John Eccles, "we have to suppose that long-term memories are somehow encoded in the neuronal connectivities of the brain. We are thus led to conjecture that the structural basis of memory lies in the enduring modification of the synapses."

However, the search for engrams and other enduring synaptic modifications through which information could be permanently stored in the brain has proved fruitless. It began in systematic fashion in the 1940s with the celebrated series of animal experiments of neurosurgeon Karl Lashley. Lashley was trying to find permanent engrams in the brain of rats by the expedient of teaching the rats specific behavioral routines and then cutting out various parts of their cortex to see where the instructions for the routines were stored. He cut out larger and larger segments of brain tissue, but found no correlation between brain area and the recall of the routine: the test animals' memory degenerated proportionately to the amount of tissue removed, but never ceased entirely. Memory seemed to pervade the whole of the rat brain. Lashley concluded that, without regard to particular nerve cells, the organism's behavior must be determined by masses of excitation within general fields of activity.

Since Lashley's finding of nonlocalized memory in rats, few neuroscientists maintain that memory is coded by localized engrams in the brain. Instead, sophisticated network-type theories are being developed, in which neurons are seen as making up diverse nets, some of which are modifiable by experience. Gerald

Edelmann, the American Nobel laureate biologist, advanced one of the best known of these "neural network" theories. His concept accounts for cognitive functions in the brain in terms of distinct neuronal groups that range anywhere from 100 to 1 million cells. Such groups respond as a unit to any signal that is conveyed to them. Each group responds to a specific subset of signal types; these are the subsets that generate attention-responses in mental processing. Since the signals select particular neuronal groups, the groups are in competition with each other in regard to their "selection" (that is, activation). For this reason Edelmann calls his theory "neural Darwinism."

The basic neuronal groups constitute the *primary repertory* of the brain: they are genetically coded and hence innate. But the groups that were once activated from the primary repertory are more likely to be selected again by the same or similar type of signals. This leads to the progressive emergence of a subset of more strongly interlinked groups, comprising the brain's *secondary repertory*. Because neuronal groups are more likely to respond to specific types of signals than to others, selective competition among them shapes the pathways of mental development. In this way mental development involves the selection of pre-existing neuronal groups by incoming signals, and the amalgamation of the groups into higher-order configurations. The mechanism of selection and group-constitution is the basis for cognitive capacity in the brain, including the discrimination of stimuli, the formation of cognitive categories, and self-recognition.

Neural network theories account for the kind of memory present in a wide variety of species, from insects to apes. The primary repertory amounts to the genetically-coded structure of the animal brain or nervous system, and the secondary repertory, since it is modifiable by experience, indicates learning capacity. Such capacity is needed in most species, since in any but the simplest organisms the rigidities of genetically-fixed behavioral routines need to be mitigated by mechanisms that enable the organism to learn from its experience—"genetic memory" alone can seldom ensure survival.

Above the level of viruses and bacteria almost every species of organism displays an experiential modification of genetically-fixed behavioral routines. Birds such as tits, for example, will hunt insects randomly if there are various species in their milieu, but if one insect species is present in larger numbers, the birds begin to hunt that species preferentially, neglecting the others. When the preferred species decreases in numbers, the birds still hunt them for a while, but then they develop another prey preference, or return to the random hunt routine. Even fish "remember" the location of the box where they were fed, though such memory lasts less than ten seconds. Memory in frogs and turtles endures for several minutes; dogs can remember a food source for several hours and sometimes days, and baboons for up to six weeks.

Our own brain performs further remarkable functions in addition to the analysis and retention—cognition and recognition—of sensory stimuli. The entire range of conscious thinking in terms of symbols, concepts and abstractions implies highly sophisticated neural information-processing, based in part on data generated by the brain itself. Feelings, intuitions, and emotional tones accompany sensory perceptions as well as abstract thought processes. However, the neurophysiological basis of higher mental functions remains essentially unknown: neuroscience is only at the beginning of a long road that may one day lead to a fuller understanding of the mind in terms of processes in the brain. The positive outlook regards not the vast seas of remaining mystery, but the ability of scientists to learn to navigate them.

IN SUMMARY . . .

More than any other sphere of nature, the human brain, the physiological seat of mind and consciousness, presents unsolved basic questions—what we already know of its workings pales in comparison with what we still do not know. Yet the understanding that has been already achieved is significant. It makes clear that the brain is not the passive camera envisaged by common sense. It is a sophisticated interpretive system that operates as an integral whole, with functions that are not decomposable to specific units, be they individual neurons or assemblies of neurons.

The brain is neither a passive system so open to the world that nothing of its own structure would be evident in perception and cognition, nor a system so closed that only its internal workings would appear to mind and consciousness. Rather, the human brain is a living part of a living system, constantly monitoring and regulating the relations between that system and the world at large. Some of that monitoring and regulating takes place in the conscious region of the brain: the phenomenon we experience as the mind. This constitutes the human brain-mind, the most accomplished information-processing system in the known world.

PART TWO

A Blurring Image

CHAPTER 5

OPEN QUESTIONS IN COSMOLOGY

A
S WE HAVE SEEN, the near shore of received knowledge is hazy in spots; the understanding conveyed by contemporary science is as yet incomplete. Of course, whether the human mind can ever achieve a complete understanding of reality is doubtful; what is certain, however, is that despite its remarkable accomplishments, present-day science has a long way to grow. Fuzzy areas persist, and conceptual black holes crop up in all the great domains of scientific inquiry. These include the universe (cosmology), matter (physics), life (biology), and mind (neurophysiology and the cognitive sciences). We review here some of the problematic issues, prior to exploring, in Part Three, the likeliest developments that could lead us to fresh horizons and a clearer vision.[1] We begin, as before, with cosmology.

THE QUESTION OF THE BIG BANG

Though the standard scenario of Big Bang theory is widely applauded, it is in trouble. There is a whole array of observations that it cannot adequately explain. This includes not only the speculative question, "What was there prior to the Big Bang—and what will be there when the process initiated by it is played out?"—but also a number of technical puzzles. For example, Big Bang theory cannot explain the "fingerprints"—the slight inhomogeneities—in the cosmic background radiation that would have led to the formation of galaxies; it fails to account for the "missing mass" in the universe (judging from the observed movement of stars within the galaxies, there is far more gravitational attraction present in the universe than matter in observed stars can account for); and it is stumped by the problem of the mechanism whereby the inflationary process of the very early universe could have been switched first *on* and then *off*.

Big Bang theory also fails to explain how it is that the texture of the background radiation, and the way stars and galaxies have evolved, is essentially the same in all directions from Earth, even in regions that are so far from one another that they could not ever have been in communication. (This is because there are cosmic regions separated by 20 billion or more light years. This distance is more than light could have traveled in all of the time—15 billion years or perhaps less—since the Big Bang.) Yet the universe has everywhere followed the same paths of evolution, obeying the same laws and manifesting the same regularities. How could this be? If the speed of light is the fastest means of interconnecting different parts of the universe, the sameness of the background radiation, and the similar ways that stars and galaxies have evolved, could only have come about by a quasi-miraculous fine-tuning of the ultra-fast expansion that followed immediately after the Big Bang.

Last but not least, mystery surrounds the age of galaxies and stars, and of the universe itself. It turns out that some galaxies are too large and too deep in space to have been created in the aftermath of the Big Bang. Four highly focused "pencil-beam" surveys have shown that there are extremely large galactic structures out at distances over a billion parsecs, with a succession of features at about 150-million parsec intervals (where each parsec equals 3.26 light years). Each of them is similar to the nearest structure, known as the Great Wall, which stretches across the sky for over 153.37 parsecs (500 million light years). These giant structures imply a far greater age for the universe than the Big Bang scenario can allow —in the estimation of some astrophysicists, more than 63 billion years.

This is vexing. *How can it be that some galaxies in the universe are older than the universe itself?*

Even if the galaxies and stars currently observed were much younger than they now appear, they might still not fit the age of the universe. That age, given by the standard scenario as 15 billion years, became controversial. Its determination depends on the precise value of the so-called Hubble constant, which gives the speed with which observed celestial objects recede from an observer on Earth.[2]

Whether *all* the structures in the cosmos originated in an explosive instability remains a matter of conjecture, regardless of whether that instability occurred 15 or 8 billion years ago. The fundamental question is not *whether* an explosive instability occurred when the standard scenario claims that it did, but whether the instability that occurred was the *first* and the *only* one. The Big Bang, after all, could also have been just one in a series of prior (and perhaps also subsequent) less all-encompassing "bangs."

Adherents of the standard scenario claim that they are not particularly disturbed by these questions. They point out that many of them are not detrimental to their model—for example, whatever triggered the Big Bang, whether an inflationary epoch or an instability in a pre-existing universe, it is not intrinsic to it. Also, questions regarding the age of the universe and its future course (recollapse, infinite expansion, or balance at the razor's edge) are part of the uncertain parameters with which cosmologists must learn to live. The standard scenario, they point out, yields nevertheless a set of meaningful interpretations and successful predictions, and these substantially outnumber its failures.[3] Some cosmologists go so far as to assert that there is no known cosmology that would give as good an account of the full range of observational and experimental evidence as the standard model.

There are, however, meaningful alternatives to BB cosmology, and they have become remarkably sophisticated lately.

The time-honored alternative to the standard scenario is the steady-state model. This was widely discussed until about 1965, from which time Big-Bang cosmology eclipsed it. In its original formulation the steady-state model maintained the basic philosophy of Einstein's theory but took care of its instabilities by specifying that matter is continually created so as to replace what is lost through expansion. Thus the universe's mean density could remain constant.

The concept of ongoing matter-creation goes back to an idea of Sir James Jeans. In 1929 he wrote, "The type of conjecture which presents itself somewhat insistently is that the centers of the nebulae are of the nature of "singular points" at which matter is pouring into our universe from some other, and entirely extraneous, dimension, so that, to a denizen of our universe, they appear as points at which matter is being continuously created." In the 1960s the noted cosmologists H.C. Arp and Sir Fred Hoyle developed this notion into the modern form of the steady-state model, replacing the idea of "matter continuously pouring into the universe from an entirely extraneous dimension" with that of matter-creation *within* it.

In the current (1993 and later) versions of their Quasi-Steady State Cosmology (QSSC), Hoyle, and his colleagues Burbidge and Narlikar show that matter-creation occurs in bursts in the strong gravitational fields associated with dense aggregates of pre-existing matter, for example, in the nuclei of galaxies.[4] According to the QSSC, underlying the general expansion of the universe is a superposed oscillation period of 40 billion years. It is at these intervals that matter-creation is concentrated, in a cycle that stretches back in time to an epoch when the scale of the universe required an oscillatory minimum. The most recent burst of major matter-creation occurred about 14 billion years ago, in reasonable agreement with the estimate of the standard scenario.

In common with other recent theories, the QSSC constitutes a multicyclic cosmology. The universe undergoes periodic creation cycles, so that matter from the current cycle coexists with matter left over from previous cycles. The galaxies in the current 14-billion-year-old universe coexist with those created earlier. Whether a galaxy is "ours" can be determined by the value of its redshift parameters: galaxies that were left over from a previous cycle recede faster relative to each other and hence have a higher-value redshift parameter than the galaxies of our own cycle of the universe.[5]

Another multicyclic cosmology is the work of Belgian Nobel laureate thermodynamicist Ilya Prigogine together with colleagues Geheniau, Gunzig, and Nardone. In a vein similar to Hoyle's cosmology, they suggest that the large-scale geometry of space-time creates a reservoir of negative energy from which gravitating matter extracts positive energy. (Negative energy is the energy required to lift a body away from the direction of its gravitational pull.) In this multicyclic cosmology gravitation plays an unsuspected role: not only does it clump together the galaxies, it is also at the root of the synthesis of matter.[6]

The "self-consistent non-Big-Bang cosmology" offered by Prigogine and collaborators outlines a perpetual mill for the creation of matter. The greater the number of particles that have been generated, the more negative energy is produced—and transferred as positive energy to the synthesis of still more particles. The quantum vacuum (of which we shall have more to say in subsequent chapters) is unstable in the presence of gravitational interaction, so that matter and vacuum form a self-generating feedback loop. The critical matter-triggered instability causes the vacuum to transit to the inflationary mode, and that mode marks the beginning of another era of matter synthesis. Thus the observed universe was not created out of an unexamined pre-existing vacuum: it arose as a new cycle within an already existing universe.

THE QUESTION OF THE FINELY TUNED CONSTANTS

Multicyclic cosmologies hold out the promise that a scientifically valid answer will be forthcoming to the perennial query, "What was there before the Big Bang—and what will be there after the matter synthesized by it has finally decayed?" But a reasonable answer, while a giant step, would still not do away with another puzzle facing contemporary cosmology. This regards the possibility of life in the cosmos.

Life, as we always knew, is possible in the universe. But we did not know that life is possible *only* in this (or in this type of) universe. Yet this appears to be the case. The conditions under which life can evolve are extremely limited; a slight variation in the basic parameters and life becomes impossible in the wide reaches of the cosmos. Luckily, the parameters of the universe are just right for life. Astrophysicists find that not only are the processes of life tuned precisely to the physical processes of the cosmos (as well they might be, since life emerged out of the physical background), but also the physical features of the cosmos are finely tuned to the conditions under which life could possibly evolve. But how could conditions that came about earlier be adapted to conditions that only came about later?

The fine-tuning of the cosmos to life involves the amount and distribution of matter in the universe, and the values of the universal forces and constants that govern the interaction of that matter. It appears that matter, though it forms but a thin precipitate in space, forms a layer of precisely the right thickness to permit the evolution of life. Were the matter-content of the universe even slightly greater than it is, the higher density of stars would create a significant probability of interstellar encounters that would knock life-carrying planets out of safe orbits. This would either freeze or vaporize whatever forms of life may have evolved on them. Moreover, if the strong force that binds the particles of a nucleus were merely a fraction *weaker* than it is, deuteron could not exist and stars such as the Sun could not shine. And if that force were slightly *stronger* than it is, the Sun and other active stars would inflate and perhaps explode.

The fine-tuning of the physical universe to the parameters of life constitutes a series of coincidences—if that is what they are. They make for a remarkable series, in which even the slightest departure from the given values would spell the end of life, or, more exactly, create conditions under which life could never have evolved in the first place. If the neutron did not outweigh the proton in the nucleus of atoms, the active lifetime of the Sun and other stars would be reduced to a few hundred years; if the electric charge of electrons and protons did not balance precisely, all configurations of matter would be unstable and the universe would consist of nothing more than radiation and a relatively uniform mixture of gases. And, if in the inflation that followed the Big Bang there had not been precise small-scale departures from large-scale regularities, there would not be galaxies and stars today—and hence no planets on which inquiring humans would seek solutions to all these puzzles.[7]

But the mass and distribution of matter in cosmic space, as well as the values of the four universal forces, were precisely such that life could evolve in the cosmos. The expansion rate of the universe and the values of the universal forces must have been determined already when this universe (or this cycle of the universe) came into being. They could hardly have been adjusted to the process they initiated purely by chance: according to Roger Penrose's calculations, achieving the kind of universe in which we actually find ourselves would call for the right choice among $10^{10^{123}}$ alternative universes. Serendipity of this magnitude strains credibility (Penrose himself speaks of a "singularity" that is beyond the laws of physics). Of course, even pure chance could create order—given sufficient time. Paul Davies estimates that the time required to achieve the level of order we now meet in the universe purely by random processes is of the order of at least $10^{10^{80}}$ years.

All of these are stupendously large numbers. We may well ask, then, whether such numbers would also apply to the different universes that may have existed before ours—or may coexist with it at present. If so, chance would be moderated by the law of large numbers: in a sufficiently large ensemble even such an improbable universe as ours becomes reasonably probable.

If we discard the hypothesis of a very large number of universes, we may have to assume that the universal constants are adjusted the way they are because only this could lead to the evolution of life—and thus to the human beings who now observe the world. Such an observer-dependent interpretation of physical reality is familiar from the philosophy of the Copenhagen school of quantum theory, and there are physicists who are willing to espouse it.

Yet it is always possible that all natural explanations will fail. Must we then face the possibility that the universe we witness is the result of purposeful design by an omnipotent master builder?

All these questions have been posed and hypotheses advanced, but no satisfactory solution has been found. Even when mitigated by the law of large numbers, chance is not a reasonable answer—it makes everything we observe, ourselves included, the plaything of a cosmic roulette. Purposeful design by a cosmic master builder overcomes this problem, but preconceived ends are even more difficult to accept for natural science than pure serendipity. And the "anthropic principle"—which tells us that the universe is the way it is because we human beings are now observing it—though widely discussed, hardly makes sense except to a particular school of quantum physicists. The puzzle persists: *how could the universe anticipate at time zero the conditions that came about 10 billion or more years later?*

Could it be that the puzzle of life and the puzzle of the Big Bang are interrelated? . . . that if we knew more about the conditions under which our universe was born, we could also find out why its constants are so remarkably tuned to the evolution of life? Perhaps . . . we shall soon see.

IN SUMMARY . . .

A reasonable explanation of why the universe is precisely so structured that life can evolve in it is still wanting. Add to this that the answers we get regarding the origins and timing of the instability that gave rise to the observed universe, though promising, are as yet largely hypothetical, and we get a picture that is fascinating, but blurred in the most intriguing spots.

We shall return to the open questions of cosmology later. But first we should set forth our review of the problems of the natural sciences with a look at the attempts of the new physicists to clear up the puzzles that beset another aspect of physical reality: the nature of matter.

NOTES

1. This review involves somewhat more attention to technical detail than was needed in the previous chapters. It is easier to describe what we believe we know than what we do *not* know: in regard to the latter we must also state why our knowledge is insufficient. However, technical details are relegated to footnotes, and even a quick perusal of the text will help the reader to gain a sense of the problems—something that he or she will find useful when we consider, in Part Three, the ways the problems could be overcome.

2. Measured by the value of the "redshift" in the frequency of the light that reaches us from stars and other objects of known luminosity (such as the star-explosions known as supernovae), the value of the Hubble constant indicates how rapidly the object moves in the opposite direction. A value of 50 indicates an age of about 15 billion years for the universe. But if the constant has a value of 80, the age of the universe comes to no more than 8 billion years. This is just what several astronomers claim, including those at the Carnegie Observatory in Pasadena and the Mount Kea Observatory in Hawaii.

3. These include the fact of the "redshift" (though no longer its exact value); the temperature of 2.7 K of the cosmic background radiation; and the proportion of hydrogen to helium in the universe (roughly $3/4$ to $1/4$).

4. Matter, assumes the model, is created in "little big bangs" of the order of about 10^{16} solar mass. It occurs through a scalar C-field of negative energy (where "C" stands for Creation), of which the value is a function of space-time. Its rate is given by the square of the time-derivative of this creation-field, averaged over the universe. Because little big bangs drive the universe's expansion, the expansion rate itself is not constant but varies with changes in the number and mass of the creation centers.

5. A good part of the cosmic background radiation itself is due to the scattering of radiation from the universe's previous cycles—the QSSC

requires 20 cosmic cycles to account for the observed properties of today's microwave background. This, however, means that the majority of photons in the universe were created not 14 or 15, nor 7 or 8, but as much as 800 billion years ago.

6. The theory maintains that there is a constant and balanced interaction between matter in the large-scale structures of the universe and the quantum vacuum—the zero-point energy field that underlies all energy and matter in the universe. In each cycle particles of matter are created in the vacuum thanks to the energy generated by particles synthesized in the previous cycle. The positive energy going into the synthesis of matter constantly and precisely compensates the negative energy generated by the curvature of space-time owing to the gravitational attraction of pre-existing matter.

7. The facts, in brief, are these:

— The expansion rate of the very early universe was precise in all directions at a rate of better than one part in 10^{40}. Yet it included small-scale departures from the large-scale uniformities—which is why galaxies, stars and planets could form in the wide reaches of space and time.

— The force of gravity is precisely of such magnitude that stars can form and exist for long enough to generate sufficient energy for life to evolve on suitable planets.

— The mass of the neutrino, if not actually zero, is small enough to have prevented the universe from collapsing soon after the Big Bang due to excessive gravitational pull.

— The value of the strong nuclear force is precisely such that hydrogen can transmute into helium and then into carbon and all the other elements indispensable to life.

— The weak nuclear force has the exact value to allow atoms to be expelled in supernovae—and thus be available in the next generation of stars to build into the more complex elements that are indispensable for life.

— The weak nuclear force also has precisely the value with respect to gravity that makes hydrogen rather than helium the dominant element in the cosmos—thus allowing stars to shine long enough, and water to form in sufficient quantities, for life to evolve on some planets.

CHAPTER 6

PARADOXES IN THE UNDERSTANDING OF MATTER

W E HAVE ALREADY noted that in the experiments and observations of contemporary physicists, matter sheds all characteristics of hard and inert substances. In many respects, matter proves to be more cloud-like than rock-like. We now take a closer look at the findings that make cloud-like matter considerably more mysterious than any cloud in the sky above us.

THE PARADOX OF NONLOCALITY

Laboratory experiments show that the smallest observable constituents of matter not only have a dual wave-corpuscle character but are also "nonlocal"—interlinked in a manner that transcends all permissible bounds of space and time. This has been known since the beginning of this century, when Young's classic double-slit experiment was first performed—which any student in physics can now duplicate. A beam is produced by an extremely weak light source, so that each photon is emitted separately (in current versions of the experiment, lasers are used for this purpose). The individually emitted photons are allowed to pass through a narrow slit in a screen. Another screen is then placed behind the first to register the photons that traverse the slit. Then, like water flowing through a small hole, the light beam made up of the photons fans out and forms a diffraction pattern. The pattern shows the undulatory face of light and is not paradoxical in itself. The paradox comes in when a second slit is opened in the screen. Then there is a superposition of two diffraction patterns, although each photon was emitted individually and has presumably traveled through only one of the two slits. Yet the waves behind the slits form a characteristic interference pattern, canceling each other when their phase difference is 180 degrees, and reinforcing each other when they are in phase. But how could the photons interfere with each other? Could they pass through both holes even though they were emitted as single particles of energy?

In a related experiment, designed by John Wheeler, photons are likewise emitted singly; they are made to travel from the emitting

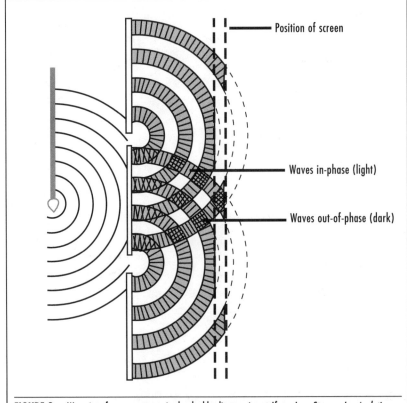

Position of screen

Waves in-phase (light)

Waves out-of-phase (dark)

FIGURE 2: Wave interference patterns in the double-slit experiment (from Jean Staune, *La révolution quantique et ses conséquences sur notre vision du monde,* 1990).

gun to a detector that clicks when a photon strikes it. A half-silvered mirror is inserted along the photon's path; this splits the beam, giving rise to the probability that one in every two photons passes through the mirror and one in every two is deflected by it. To confirm this probability, photon counters that click when hit by a photon are placed both behind the half-silvered mirror and at right angles to it. The expectation is that on average one in two photons will travel by one route, and the other by the second route. This is confirmed by the results: the two counters register a roughly equal number of clicks—and hence of photons. When a second half-silvered mirror is inserted in the path of the photons that were undeflected by the first mirror, one would still expect to hear an equal number of clicks at the two counters: the individually emitted

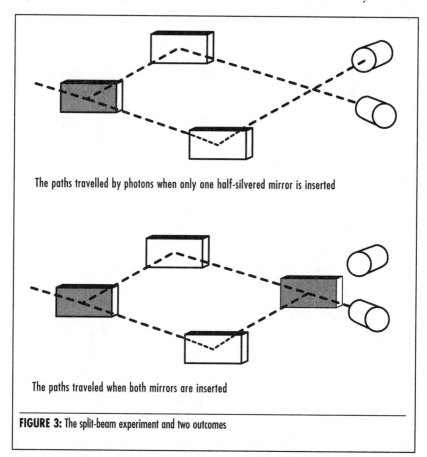

The paths travelled by photons when only one half-silvered mirror is inserted

The paths traveled when both mirrors are inserted

FIGURE 3: The split-beam experiment and two outcomes

photons would merely have exchanged destinations. But as Figure 3 shows this expectation is not borne out by the experiment. Only one of the two counters clicks, never the other. All the photons arrive at one and the same destination!

It appears that the kind of interference that was noted in the double-slit experiment recurs. Above one of the mirrors the interference is destructive (since the phase difference between the photons is 180 degrees) so that the photons, as waves, cancel each other. Below the other mirror the interference is constructive: the wave-phase of the photons is the same and they reinforce one another. *Could it be that the photons, emitted as individual particles, interfere with each other as waves?*

Photons that interfere with each other when emitted moments before in the laboratory interfere with each other also when emitted at considerable time intervals in nature. The "cosmological" version of Wheeler's experiment bears this out. In this experiment the photons are emitted not by an artificial light source but by a distant star. In one case the photons of the light beam emitted by the double quasar known as 0957+516A,B were tested. This distant quasi-stellar object is believed to be one star rather than two, the double image being due to the deflection of its light by an intervening galaxy situated about a quarter of the distance from Earth. (The presence of mass, according to Einstein, curves space-time and hence also the path of the light beams that propagate in it.) The deflection due to this "gravitational lens" action is large enough to bring together two light rays emitted billions of years ago. Because of the additional distance traveled, the photons that were deflected by the intervening galaxy have been on the way 50,000 years longer than those that came by the more direct route. But, although originating billions of years ago and arriving with an interval of 50,000 years between them, the photons interfere with each other just as if they had been emitted seconds apart in the laboratory.

The interaction of particles, even at large distances from one another, is quasi-instantaneous. This paradoxical facet of nonlocality was confirmed in the testing of the so-called "EPR" experiment, originally advanced by Einstein with colleagues Podolsky and Rosen.

The experiment concerns a pair of particles that are in identical quantum states but propagate in opposite directions. A measurement of position is made on one of them and, since the particles are in identical states, these results are then used to predict the corresponding state of the other. Then on the second particle another property is measured, in this case, momentum. If this should prove possible, *both* the momentum *and* the position of the second particle would become known. This, however, is forbidden by Heisenberg's indeterminacy principle which tells us that when one parameter of the state of a particle is measured, the other becomes entirely unmeasurable. Einstein expected that this experiment would show that the indeterminacy principle was not an intrinsic feature of nature, only a consequence of the act of measurement.

Though the EPR experiment was suggested in 1935, it was only in 1982 that laboratory equipment could be built to test it. The test, performed in France by Alain Aspect and collaborators, showed that Heisenberg's principle is not violated—but not in the way expected. It turned out that, despite the spatial distance, the act of measurement on one particle has a measurable effect on the other. More precisely, the indeterministic quantum state of the second particle resolves into the deterministic state typical of observed particles as soon as the first particle is measured—the wavefunction of both particles "collapses" at the same time. Much as in the split-beam experiment, two particles, as long as they originated in the same quantum state, prove to be correlated with each other even when they are separated in space. And this correlation occurs quasi-instantaneously; sophisticated instrumentation shows that it is faster than the speed of light.

The phenomenon of nonlocality prompted the invention of other famous thought experiments, including that which became widely known as "Schrödinger's Cat." German physicist Erwin Schrödinger proposed that we take a cat and place it in a sealed container. We then set up a device which, entirely randomly, either does or does not spray a poisonous gas into the container. Thus when we open the   container the cat is either dead or alive. Common sense would suggest that either the cat dies when the gas is emitted—if it is emitted—or it remains alive throughout the time it spends in the container. But this state of affairs is forbidden in quantum theory. As long as the container is sealed, there is a probabilistic superposition of states: the cat must be *both* alive *and* dead. It is only when the container is opened that the two probabilities (which stand for the cat's wave-function) collapse into one.

A similar thought experiment was proposed by the French scientist Prince Louis de Broglie. Here we place an electron rather than a cat in a sealed container. We divide the container, which is in Paris, and ship one part to Tokyo and the other to New York. This time common sense would dictate that if we open the half which is in New York and find the electron, then the electron must have already been in that half already when the container was shipped from Paris. But this state of affairs, like that which decides whether Schrödinger's cat is alive or dead, is forbidden. Each half-container must have a non-zero probability of harboring the electron. Then, the instant one of the halves is opened in New York the electron's location is decided. The wave-packet defining the probability of the electron's presence is reduced also in Tokyo.

In the standard version of the experiments—and also in the thought experiments that gave rise to them—particles are assumed to remain coordinated as long as they were once "one"—that is, as long as they originated in an identical quantum state. It turns out, however, that particles can be instantly correlated even if they were not associated before. German physicist Gerhard Hegerfeldt of the University of Göttingen discovered this in 1995, when he reviewed Enrico Fermi's 1932 calculations concerning the interaction of two atoms, one of which is in an "excited" state. Fermi wanted to know how an atom's moving down from an excited to a "ground" state would affect another atom. It is known that as the atom radiates off the extra energy of its state of excitation, that radiation will excite a second atom to the corresponding extent (a principle that underlies the workings of laser). Fermi naturally assumed that the effect will be delayed just by the time it takes for the radiated energy to travel from the first atom to the second. But Hegerfeldt's revision of the calculations showed that the second atom can be excited the very instant the first one decays. When an electron in the first atom moves down a given energy level, its counterpart electron in the second atom moves instantly up by the same amount. It appears that the wave-function of the electron in the excited atom overlaps that of the electron in the atom subjected to the excitation. The two electrons are correlated in much the same way as the originally identical and then separated electrons are correlated in the EPR experiment. *Could it be, then, that the denizens of the quantum world are constantly and almost instantly correlated with each other?*

OTHER PARADOXICAL CORRELATIONS

There are other forms of mysterious correlation among the basic entities of the quantum world. They come to light in superconductors and superfluids. The puzzling feature of these forms of correlation is similar to those we have noted above: it is instant and it does not involve any known force or medium.

When various pure metals and alloys are supercooled to within a few degrees of absolute zero, their electrical resistance vanishes. The substances become superconductors: an electrical current passing through them is transported entirely without friction. This phenomenon was discovered by Kamerlingh Onnes in 1911, and its details, together with those of superfluidity (the lack of viscosity in a supercooled liquid such as helium), came to light in subsequent decades in research in low-temperature physics. It turns out that as a metal or alloy is cooled to a critical temperature, electrons flow through it in a fully coherent manner. A similar phenomenon occurs in superfluids. Previously randomly colliding molecules cohere into a single quantum entity without apparent viscosity; consequently such a fluid can flow through capillaries and cracks without resistance. In both cases a highly cohesive quantum state is generated. The Schrödinger wave-function of the motion of all the electrons in a current, and of all the particles that make up the molecules of a fluid, assumes one and the same form.[1]

It appears that the electrons in a superconductor, and the particles that make up the molecules of a superfluid, are precisely and continuously correlated with each other. Yet there is no dynamical force connecting them.

Recent research discloses that the coherence induced by instant correlations between superconductors is more widespread than it was originally believed. Brian Josephson, who received the Nobel Prize for his discovery, found such correlations among superconductors even when they are separated by finite distances. The curious "Josephson effect" also occurs at normal temperatures. According to Italian biochemist Emilio Del Giudice and collaborators, the correlation obtains in neighboring systems of material entities, whether they are particles, atoms, or molecules. A pair of nearby cells can act as a Josephson junction, and a set of identical cells can create an entire array of such junctions with the phase of the cells' vibrations locked in. Since coherence between individual cells produces coherence among entire cellular assemblies, this effect may be a major factor in ensuring the overall integrity of the living organism. *But if they are not linked by an energy or signal-carrying medium, how does one particle—or molecule or cell—"know" the state of another?*

Nonenergetic correlation among particles also occurs within the electron shells of atoms. The effect in question—known as Pauli's exclusion principle—was described by Wolfgang Pauli as far back as in 1925. It involves atoms orbiting the same atomic nucleus (in the case of complex molecules, the same set of nuclei). To grasp this phenomenon we should recall that the composition of the atom's nucleus determines the energy levels that can be accommodated in its surrounding shells, but the energies of the nucleus do not determine the *distribution* of the energy levels within the shells. That is determined only by the correlation among the electrons themselves—a correlation based on the mutual exclusion of electrons in accordance with Pauli's exclusion principle. That principle shows that electrons in an atom always assume states with an antisymmetric wave-function.[2] To satisfy this penchant for antisymmetry, electrons surrounding atomic nuclei would have to know the wave-function of each other's quantum states. But how they could do so is not clear. The exclusion principle calls for precise correlation between electrons without allowing that a dynamical force would be exchanged among them. In the same way as two particles in the EPR experiment and two photons in the split-beam experiment are "informed" of each other's quantum state without the exchange of manifest energy, energetically unlinked electrons in an atom, molecule or metal are mutually informed of each other's quantum states.

Though its dynamics are puzzling, Pauli's principle explains why matter configures into increasingly complex structures in the cosmos. It is because around atomic nuclei, electrons are constrained to occupy unique quantum states and so produce differentiated structures, rather than forming larger and larger blobs of matter-energy. Atoms with differentiated structures can combine with one another according to the specific properties of their valence. As a result of this, complex systems of atoms can build up in the course of time.

For matter to build toward higher levels of structure and complexity, further electrons must enter the energy shells of the existing nuclei from time to time. This requires that in the pertinent reactions the energy levels of the contributing and receiving nuclei are harmonized. In the universe at large, such harmonization seems highly improbable. Yet it must have occurred with remarkable frequency, since in many parts of the cosmos matter has built into comparatively complex structures, such as heavy elements and complex molecules integrating several (and conceivably a very large number of) atoms.

The key to higher-order atomic structures in the universe is the presence of carbon. As we have noted in Chapter One, in the early universe hydrogen nuclei were the first to be synthesized. Subsequent reactions fused some of the hydrogen nuclei into the more complex nuclei of helium. But both hydrogen and helium are inert, and in the expanding and cooling universe the amount of energy required to make them combine into heavier elements was not available. Still more complex elements could only have evolved if there was enough carbon on hand to catalyze the reactions that would make hydrogen and helium combine further, into heavier nuclei. Enough carbon was indeed available, but why it was is due to an astonishing coincidence. The energy levels of carbon, beryllium, helium, and oxygen were in harmony.[3]

Despite the low probability that the energy levels of carbon, helium, beryllium and oxygen would be fine-tuned to the required degree, it was the case. Nature exhibits a puzzling fine-tuning of the energy levels of four different elements. It is due to this fact (together with the already noted fine-tuning of the universal constants) that the universe manifests things and events that are more interesting than a randomly swirling gas of hydrogen and helium.

We may well ask, however—*is the harmonization of the frequencies of four different elements pure coincidence?*

The curious phenomena described here are so many variations on a basic principle, accepted by quantum physicists as a fundamental feature of the physical universe. This is nonlocality. Closer acquaintance shows it to be a variant of the old but controversial concept of "action-at-a-distance." The finding to be explained is that an event at point A affects another event at point B even though the two points are not next to (that is, contiguous with) each other. The commonsense explanation is that something—a propagating force, or else a continuous surface or medium—carries the cause from point A until it issues in the effect at point B.

Modern quantum mechanics does not go along with common sense—it does not assume a connecting force or medium. Instead, it postulates a correlation between events at A and B. This correlation is quite definite. In the EPR experiment, for example, measuring a spin component of a particle at point A (let us call it particle A) "causes" in a quite precise way the collapse of the wave-function of particle B. Particle B always "collapses" into a state with the opposite spin component. If we measure different things in A, correspondingly different collapse-effects surface in B.[4]

This finding is entirely contrary to the commonsense concept, where there is either an interconnecting force or medium between separated things or events, or there cannot be any transmission of cause and effect between them. If A and B are separated, they must be truly separate. Experiments show, however, that this is not the case. Separated quantum events remain in some mysterious way "together." Yet, even though this appears to be a basic feature of the physical world, quantum theory does not postulate a continuous force or medium that would connect the "separately-together" events. Why not? For one thing, the transmission of the effect from one event to the other turns out to be quasi-instantaneous—far faster than the speed of light—and that violates the principle of relativity. No ordinary force could be involved here. For another, quantum physicists are apprehensive about continuous mediums in the universe—they fear a return of the discredited 19th-century notion of a luminiferous ether. Thus they accept a typically Alice in Wonderland situation: there is the observed effect (the grin of the Cheshire Cat) but there is nothing that would "carry" that effect (no Cat itself).

But are the current assumptions of quantum theory the final word about the nature of physical reality? Some scientists think that they may not be. Quantum mechanics, as Einstein had already said, is in some essential respects still incomplete. Today even quantum theorists, such as J.C. Polkinghorne, note that physics may not yet have succeeded in taking in fully what quantum mechanical nonlocality implies about the nature of the world.

IN SUMMARY . . .

Matter, as science has come to know it, has little if anything to do with the classic notion of inert bits of rock, or even with mathematical mass-points. (From our point of view this is extremely fortunate: inert chunks of matter could never have built into the intricate arrays of order and organization that are the precondition of the higher forms of life.) The particles that possess matter-like properties are complex dynamic entities with a number of paradoxical characteristics. They exhibit the phenomenon of nonlocality, as well as dual wave-like and corpuscle-like properties. They manifest a penchant for cooperation, adopting highly coherent states without known energetic or informational links between them. And their energy levels are harmonized so precisely that they can cohere into the complex atomic—and molecular and supramolecular—structures that we now observe in the universe.

NOTES

1. Normally, when an electrical current passes through a metal, it produces a drift in the electron gas—the electrons are scattered from vibrating atoms in the lattice structure of the metal. This retards the flow of electrons through the lattice and produces the friction that heats the metal: hence the phenomenon of electrical resistance. However, when the metal is supercooled, the vibrations of the atoms are reduced and the resistance of the metal is lowered. Since even near the absolute limit of temperature zero-point energies should keep the lattice vibrating, electrical resistance should actually be present even when metals or alloys are cooled to within a few degrees of absolute zero. Yet at these temperatures resistance vanishes entirely: in a ring built of a superconductor, an electrical current, once induced, keeps flowing indefinitely.

2. This rule means that each electron in an atom occupies a different orbit. Pauli's principle is a consequence of the formalisms of quantum mechanics when we hold that the only solutions of the Schrödinger equation that are physically possible are those that are antisymmetric under permutation of the electron's coordinates. In everyday language this means that as we add further electrons to the atom, the incoming electron will not occupy an orbit that is already taken: it is excluded to another orbit, of which the wave-function has the opposite symmetry.

3. The synthesis of carbon calls for a series of reactions that begins with the helium + helium reaction: this produces a nucleus of beryllium. The resulting beryllium nucleus is an unstable isotope: it disintegrates into helium almost as soon as it is created. In order to produce carbon, rather than disintegrating into helium, beryllium would have to enter into reaction with it. This reaction, though it is highly improbable, does take place. It does so because it is a "resonance reaction," where the combined energy of the beryllium and helium nuclei (7,370 million electron volts [MeV]) is just slightly less than the energy of carbon, the product of the reaction (7,656 MeV).

It is not assured, however, that the carbon produced in this reaction would survive: a further reaction—carbon + helium—would reduce it to oxygen. But it so happens that the carbon + helium reaction is not favored by nature: the energy level of the reaction-product oxygen (7.1187 MeV) is below the energy level of the reactants carbon + helium (7.1616 MeV). As a result the nucleus of oxygen is relatively stable, and both carbon and oxygen are available in sufficient quantities to become elements in still more complex elements, including those that are the basis of life.

4. Different effects are produced according to just what component of A is being measured. For example, if the measured spin component is one which is "up" along the z-axis, then B will be in a state where the spin-component is "down" on that axis. And when in A the up-spin along the x-axis is measured, then B proves to be in a state of equal superposition of the "up" and "down" states along the z-axis.

CHAPTER 7

PUZZLES IN THE PHENOMENA OF LIFE

THE NATURE OF LIFE is far too great a question to be definitively answered by any theory based on observation and experiment; findings can always come to light that question old tenets and bring new, and sometimes quite radical, insights. Indeed, in recent years there has been no dearth of radical findings in the sciences of life; as a result the standard account has undergone considerable revision. Yet the basic concept has not been called into question: it is still maintained that life has evolved in the biosphere from non-life, in consequence of favorable physical and chemical conditions and constant irradiation by solar energy. But just *how* life has evolved has become a bone of contention.

THE PUZZLE OF THE LEAP

How life would have evolved—that is, how new species would have emerged from the old—has been described by Darwin. According to his classical theory, natural selection acts on random mutations. The latter are "typing mistakes" in the repetition of the genetic code of the parent in the offspring; and such mistakes are produced by all species at a more or less constant rate. Most of the mutants due to chance variations of the germline are faulty in some respect and will be eliminated by natural selection. However, random mutations occasionally hit upon a genetic combination that renders the offspring more rather than less fit than the parent to live and to reproduce. Such an individual transmits its mutant genes to successive generations, and in time the comparatively numerous offspring of these generations displace the previously dominant species. The range of possible genetic variations is limited only by the comparative fitness of the mutants for life and reproduction in their particular habitat.

But this account of the classical Darwinian theory fails to mesh with the evidence. Paleobiologists, experts at sifting the fossil record, contest that natural selection would produce a process that is gradual and continuous. The fossil record suggests that evolution could have missed several links in a presumably continuous chain: new species have appeared suddenly, without gradual transitions having led up to them.

It seems that the "phylogenetic incrementalism" of classical Darwinism is not correct. Darwin himself may have professed such incrementalism more from a conservative predisposition than on the basis of scientific evidence. He followed Linnaeus in affirming that *natura non facit saltum*—nature does not make leaps. Sudden leaps in nature resemble revolutions in human society, and the dominant mentality of Darwin's time extolled piecemeal adjustments while it abhorred wholesale transformations.

Darwin, biographical researches show, was very likely influenced by the dominant mentality of his time. Yet nature disregards the dispositions of 19th-century Englishmen and progresses by sudden leaps and radical transformations nevertheless. In 1972, almost 120 years after the original publication of *The Origin of Species,* Stephen Jay Gould and Niles Eldredge came out with an influential study that introduced the leap into neo-Darwinian theory.

In the Gould–Eldredge theory of punctuated equilibrium— where "equilibrium" refers to a dynamic balance between species and environment—evolutionary processes concern entire species rather than individual reproducers and survivors. Evolution occurs when the dominant population within a clade (a set of species sharing a similar adaptive plan) is destabilized in its milieu and other species or subspecies that had emerged haphazardly on the periphery break through the cycles of dominance. At that point the stasis of the epoch is broken, and there is an evolutionary leap from the formerly dominant species, threatened with extinction, to peripheral species or subspecies. The process is relatively sudden: speciation punctuates long periods during which species persist basically unchanged. This means that as long as a species persists, it remains relatively unchanged: its genetic information pool is handed down more or less intact to succeeding generations. At the end of its collective life span it does not transform into another species, but dies out and is replaced by better-adapted species.

This is the surprising finding suggested by the fossil record. It testifies that living species did not evolve in a continuous and piecemeal manner: there were periods of millions of years during which the existing species did not change in any significant way. New species tended to burst on the scene within relatively short time periods—somewhere between 5,000 and 50,000 years.

Recent evolution theory recognizes both long periods of stasis and short periods of sudden, and in detail unpredictable, transformation. The classical Darwinian mechanism of adaptation functions only during the former: it molds species to their given environment. When that adaptive process is interrupted—perhaps by changes in the environment—it breaks down and a transformatory process takes over. The system built of a species and its environment enters a chaos-like state where the smallest fluctuation can create a decisive choice between the alternative pathways along which it could evolve. This chaotic, nonclassically determined process is known as a system "bifurcation." (The name indicates that the evolutionary path or trajectory of the system no longer continues unchanged: it "bifurcates" into a new mode.) The forces and constraints that define the system are reconfigured, so that a fundamentally different system emerges.

The element of indeterminacy in the bifurcation process applies only to individual species and not to the evolutionary process as a whole. Despite the unpredictability of the evolution (or extinction) of a given species, there is an overall predictability associated with the course of evolution on this planet. As the fossil record shows—and as we have already seen in Chapter Three—biological evolution moves in a preferred direction. In the course of evolutionary time, many species grow from single-celled protobionts and simple algae into larger and more complex organisms. The biosphere is now populated with species at all levels of size and complexity. The more evolved among them moved sequentially, though by no means smoothly and continuously, from microscopic size and comparative simplicity of structure to larger size and higher complexity.

THE PUZZLE OF CHANCE

The standard Darwinian account perceives a twofold intervention of chance in the evolutionary process: first, by bringing forth mutations in the genome, and then by exposing the mutant organisms to an environment in which they could survive. Some investigators, such as Richard Dawkins, appear perfectly satisfied with this twofold element of chance. According to Dawkins, the evolution of the gene pool occurs through trial and error, so that the evolution of living species resembles the work of a "blind watchmaker." Given sufficient time, evolution's trials and errors will generate all the living forms that have ever populated the biosphere.

Other scientists are less convinced. An outspoken critic of Darwinism, Michael Denton asked whether random processes could have constructed an evolutionary sequence of which even a basic element, such as a protein or a gene, is complex beyond human capacities. Can one account statistically for the chance emergence of systems of truly great complexity, such as the mammalian brain when, if specifically organized, just 1 percent of the connections in such a brain would be larger than the connections in the world's entire communications network? Denton concluded that chance mutations acted on by natural selection could well account for variations *within* given species, but hardly for successive variations *among* them.

Chance is a problem already because the assembly of even a primitive self-replicating prokaryote involves building a DNA double helix of some 100,000 nucleotides, with each nucleotide containing an exact arrangement of 30 to 50 atoms, together with a bi-layered skin and proteins that enable the cell to take in food. This construction requires an entire series of reactions, where each reaction brings about a further decrease in the internal entropy of the system. Sir Fred Hoyle pointed out that this process's occurring purely by chance is about as likely as a hurricane's blowing through a scrap yard assembling a working aeroplane.

Some years earlier, Konrad Lorenz came to a similar conclusion. While it is formally correct, he said, to assert that the principles of chance mutation and natural selection play a role in evolution, by itself this cannot explain the facts. Mutations and natural selection may account for variations within given species, but even the nearly 4 billion years that were available on this planet for biological evolution could not have sufficed for chance processes to generate today's complex and ordered organisms from their protozoic ancestors. *But if not chance, then what has been responsible for the evolution of order and complexity in the living world?*

The problem is not new. In the 1950s mathematician Hermann Weyl noted that, since each of the molecules on which life is based consists of something like a million atoms, the number of possible atomic combinations is astronomical. On the other hand, the number of combinations that could create viable genes is relatively limited. Thus the probability that such combinations would occur through random processes is negligible. A more likely solution, said Weyl, is that some sort of selective process has been taking place, probing different possibilities and gradually groping its way from simple to complicated structures. Weyl himself was of the opinion that "immaterial factors"—in the nature of images, ideas, or building plans—are likely to be involved in the evolution of life.

Weyl's speculations have not been accepted by the scientific community: scientists believe that nature creates its own designs instead of receiving them ready-made. Yet some sort of design does seem to be present. French biologist Jean Dorst, though reluctant to admit teleology (causation by an event that is as yet in the future), was forced to conclude that there is, after all, a design inscribed in nature: a design observed in the balance between different species as well as in some extraordinary adaptations, such as between plants and insects. These, he said, go far beyond the facts explicable by Darwinian theory. Etienne Wolff, in turn, spoke of "orientation" in evolution. There were ten or more precursors of the family of mammals between the end of the primary and the beginning of the secondary era, but only one among them gave rise to today's mammals. There were also many types of species that tried taking to the air, including dinosaurs, pterosaurs and reptiles, even the archaeopteryx, but only one variety succeeded. At every level of the hierarchy of animals there appears to be a tendency in evolution to produce something new, more adapted and more complex. It is evident, said Wolff, that a chance process would not have evolved the kind of order and consistency that now meets the eye. If evolution had been at the mercy of chance, its course would have been entirely different.

Chance appears to be an insufficient explanation also in regard to the consistency that living species exhibit relative to each other. The wings of birds and bats, for example, are homologous with the flippers of the phylogenetically entirely unrelated seals and with the forelimbs of the equally unrelated amphibians, reptiles, and vertebrates. While the size and shape of the bones show great variation, the bones themselves are similarly positioned, both in relation to each other and to the rest of the body. Diverse species also exhibit common orders in regard to the position of the heart and the nervous system: in endoskeletal species the nervous

system is in the dorsal (back) position and the heart in the ventral
(front) position, while in exoskeletal species the positions are
precisely reversed.

It is no less remarkable that some highly specific anatomical
features should be shared by species of plants and animals that
evolved in widely different locations and had entirely different evolutionary histories. The more than 250,000 species of higher plants
found in almost every part of the world show only three basic
forms of the distribution of leaves around the stem—and a single
form (the spiral) accounts for 80 percent of all of them. British
biologist Brian Goodwin made a compelling case that this is
not due to the accidents of genetic mutation and natural selection.
Rather, he notes, there is an inherent pattern to life that makes
it intelligible at a much deeper level than functional utility and
historical accident.

Animal species, too, though they show great variation in complexity at all levels—the level of genes as well as of proteins—have
remarkably similar structural forms or morphologies. Biologists
note that diverse species can construct nearly identical *phenomes*
(phenomenological real organisms) on the basis of vastly different
genomes. Even more astonishing is the reverse case: when the
same elements of a genome recur in widely different phenomes.
Closely similar, even identical, genes have been found in species
that had evolved entirely independently of one another. The most
remarkable case is that of the eye.

Swiss biologist Walter Gehring and his collaborators discovered
that the three dozen or so eyes that occur in the living realm
have a common origin. The reticular eye of the fly and the retina-
covered visual organ of mice and men are derived from the same
basic pattern; indeed, they are coded by one and the same "master
control gene." The genetic mechanism of the eye is interchange-
able among widely differing species—the "eye-gene" of the mouse
will induce the growth of an eye on the fly.

The information coded by this master control gene must have

evolved over 500 million years ago. Since then, it has been adopted and adapted by nearly 40 phylogenetically distinct types of insects and animals. But how did these different species acquire one and the same information for building their visual organ? *Could it be that they accessed the information from each other—or from some archetypal form or pattern in nature?*

The puzzle of chance extends to the very beginnings of life on Earth. Complex structures have appeared on this planet within astonishingly brief periods of time. The oldest rocks date from about 4 billion years, while the earliest and already highly complex forms of life (blue-green algae and bacteria) are, as we know, over 3.5 billion years old. How this level of complexity could have emerged within the relatively short time of about 500 million years lacks a satisfactory answer. Chance alone cannot account for the facts: a random mixing of a molecular soup would have taken incomparably longer to produce these structures. In their search for a reasonable explanation, theoreticians have been known to indulge in flights of imagination. In the 19th century, Lord Kelvin maintained that life is likely to have been imported to Earth "ready-made" from elsewhere in the universe. More recently DNA-decoder Sir Francis Crick revived this idea.

The improbability that evolution would have proceeded by random trial and error is compounded by the fact that the environment in which biological species evolve is far from constant. What was once a suitable habitat may become less suitable in time, and may even threaten the survival of some species. In order to remain viable in a different milieu, living species have to modify their adaptive plan. How such modification could be accomplished on Darwinian premises is a mystery. If a species proceeded by random mutations, it would risk maladaptation—and possibly extinction—before it could reach a state of adaptation to the alternative niche that had become available to it.

The kind of changes in DNA that could assure the viability of a species in a changed habitat are not likely to have been produced

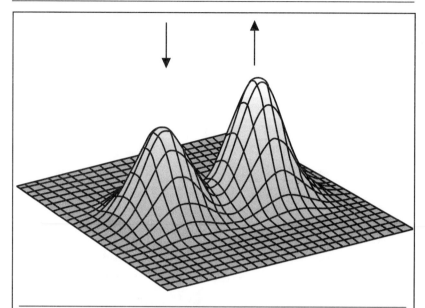

FIGURE 4: Topographical models picture the problem of viable mutations as crossing from one "peak of adaptation" to another within a so-called adaptive landscape. Mutations move a species along the landscape in two dimensions. The third dimension is the level of adaptation of a species to its niche. Mutations generally drive a species up an adaptive peak, since maladapted mutations die out and well-adapted ones are reproduced. The problem comes about as a niche becomes less viable, or disappears altogether (sinks or disappears as an adaptive peak). The species that were adapted to it must adapt to another niche. But, if mutations that diminish the level of adaptation to the current hill are weeded out by natural selection, this is a problem. How to create those massively systematic mutations that could propel the species to the foot of another, more stable or perhaps rising, adaptive peak? Once there, stepwise mutations will drive it up its slope. But getting there is a puzzle.

by purely random modifications of the genetic pool. This is because it is not enough for mutations to produce one or a few positive changes in the organism; they must produce a full set. The evolution of feathers, for example, does not produce a reptile that can fly: radical changes in bone structure and musculature are also required, along with a faster metabolism to power sustained flight. Each innovation by itself is not likely to offer evolutionary advantage; on the contrary, it is likely to make the organism less fit than the standard form from which it departs. If so, it would soon be eliminated by natural selection. A random stepwise elaboration of the genetic code of a species would not ordinarily produce viable results.

French-born Harvard biologist M. Schutzenberger noted that one would need an almost blind faith in Darwinian theory to believe that chance alone could have produced in the line of birds all the modifications needed to make them high-performing flying machines, or that random mutations would have led to the line of mammals after the extinction of the dinosaurs—given that mammals are a long way from dinosaurs along the axis that leads from fish to reptiles. Giuseppe Sermonti was of the same opinion: it is hardly credible, he said, that small random mutations and natural selection could have produced a dinosaur from an amoeba. Life, it seems, does not evolve by piecemeal improvements but by occasional, but then massive and revolutionary, innovations.

In regard to the tempo and the mode of evolution the classical Darwinian account is falling by the wayside. The recognition that dawns is that random mutations exposed to the test of natural selection are not likely to have brought forth the complex and highly adapted species indicated by the fossil record—not, at least, within the known time frames. It is said that the mills of the gods grind extremely fine and very slow—but in regard to the evolution of life they could not have ground quite as slow as classical Darwinism would require.

Biologists now admit that truly significant evolutionary events, such as the emergence of new species, cannot be adequately explained on the assumption that macroevolution is the sum of the set of randomly produced and naturally selected microevolutionary modifications. Not only could new species not have arisen by the stepwise modification of previous species, the fossil record itself testifies against such a "gradualistic" thesis. Major novelties did not result from the accumulation of minor changes: the variational possibilities were too vast, and the observed leaps between the species too great to have permitted random variations in the germline to produce the observed course of evolution. *But, then, how did the species that were threatened by unfavorable changes in their environment manage to survive? Why did they not die out, leaving the biosphere populated mainly by algae and bacteria?*

THE PUZZLES OF MORPHOGENESIS

Modern genetics faces a further challenge when it comes to processes of morphogenesis (the regeneration of the organism, usually through reproduction). Single-celled organisms reproduce by division, transferring the DNA of their chromosomes to new cells by splitting. However, more complex species reproduce from their reproductive cells. Presumably, each of these cells has a set of instructions that suffices to build the entire organism. But does it?

The fact that species breed true—that out of a chicken egg comes a chicken and not a pheasant—calls for explanation. The standard explanation furnished by genetics is that the genetic code in the reproductive cells of a species contains the blueprint for the whole organism. This, however, is not free of difficulties. To begin with, as we have just noted, the genetic code is often closely similar among widely divergent species, and divergent among relatively similar ones. The DNA in the chromosome of the chimpanzee is 98.4 percent the same as that of humans, while amphibians that share many morphological features turn out to have widely differing genetic information. Then the basic assumption of standard genetics—that a given gene has a determinate and uniquely correlated effect on the organism—has proven to be a considerable exaggeration. Genes alone do not determine malfunctions and diseases, not even processes of development and aging. For the most part, complex organic processes have more than a genetic basis. Though there are genes that control entire strings of complex processes, development usually involves an interaction between many genes, gene products, and environmental (so-called epigenetic) factors. For a disease such as cancer or heart disease up to a thousand genes may be involved, and their interaction is fundamentally influenced by effects in different environments. As population geneticists well know, a "precipitating" environment is required for the manifestation of a disease across the full range of its genetic variation.

Organic functioning, it appears, is not a direct—so-called linear —consequence of the information encoded in the genes. Instead, such functioning involves a complex nonlinear process that has many elements in common with the dynamics of chaos.

This finding is especially evident in the staggeringly complex processes involved in embryogenesis. In the case of mammalian species, the development of the embryo requires the ordered unfolding of myriad dynamic pathways in the womb, involving the coordinated interaction of billions of dividing cells. If this process were entirely coded by genes, the genetic program would have to be miraculously complete and detailed. It would also have to be flexible enough to ensure the differentiation and organization of a large number of dynamic pathways under a potentially wide range of conditions. Yet the genetic code is the same for every cell in the embryo. It is unlikely that this code alone could conduct and coordinate the full range of developmental interactions.

French Nobel laureate biologist François Jacob admitted that very little is known about regulatory processes in embryonic development. Aside from as yet vague notions of epigenetic landscapes and biological fields, the only logic that biologists really master is linear and one-dimensional. If molecular biology was able to develop rapidly, said Jacob, it was largely because information in microbiology happens to be determined by linear sequences of building blocks. And so everything turned out to be one-dimensionally linear: the genetic message, the relations between the primary structures, the logic of heredity, and so on. Yet, in the development of an embryo, the world is no longer linear. The one-dimensional sequence of bases in the genes determines the production of two-dimensional cell layers that fold in precise ways to produce the three-dimensional tissues and organs that give the organism its particular morphological shape and physiological properties. According to Jacob, how this occurs is still a complete mystery. The principles of the regulatory circuits involved in embryonic development are not known. For example, while the molecular anatomy of a human hand is understood in some detail, almost nothing is known about how the human organism instructs itself to build that hand.

It appears that the organism can both build, and to some extent rebuild, its damaged parts with great precision. For example, when a finger of the human hand is amputated above the first joint and the wound is not sealed surgically with skin, the tip of the finger can be regenerated. Astonishingly enough, the regrown fingertip is complete down to the finest detail, reproducing even the individual's unique fingerprints.

Darwinian theory maintains that programs of organic self-repair must have been naturally selected in the course of the prior history of the species. Random mutations that happened to hit upon modifications that improved the capacity of the organism to repair the kind of damages that kept befalling it had survival value; the mutants that possessed them had more numerous and more viable offspring than the rest—in the end they became the dominant population. The proposition sounds reasonable, yet it fails to mesh with the evidence. Many organisms turn out to possess programs of self-repair that could not have been naturally selected, for the kind of damage they repair is not likely to have occurred in the entire natural history of the species.

The mysteriously unprecedented programs come to light when simple organisms are subjected to sophisticated manipulations in the laboratory. Obviously, these could not have occurred in nature. For example, when experimenters cut up a common marine sponge—a true multicellular organism consisting of various specialized cells with coordinated functions—and squeeze its cells through a sieve fine enough to break apart all connections between them, the set of seemingly disconnected cells is able to reassemble into the full organism. The sea urchin—a more complex organism, complete with digestive tracts, vascular systems, tube-like feet for locomotion and a ring of plates surrounding the skeletal scaffold— can perform similar self-repair. When it is deprived of the calcium required for its skeletal frame, its parts disassemble and the sea urchin dissolves into a mass of separate cells. But when the required level of calcium is reintroduced, the cells reorganize themselves and reconstitute the full sea urchin.

The frog is capable of a comparable feat. When one takes a fertilized ovum of this amphibian and places it in a centrifuge, the strong G-force mixes up the components of its cellular structure. Yet the ovum can still develop into a normal frog. One can divide the egg of a dragonfly in two and destroy one of the halves: the other can nevertheless develop into a complete dragonfly. As any child knows, a humble flatworm can be cut into several pieces, and each segment can grow back into a complete worm. One can cut off the leg of a newt and the newt—unlike the otherwise closely similar frog—will grow a new leg. It will even regenerate the lens of its eye: when an eye is surgically removed, the tissues at the edge of the iris reassemble into a new lens. Faced with these findings, we cannot help but wonder: *Seeing that programs that repair artificial damages inflicted by scientific curiosity in the laboratory could not have been part of the survival kit of living organisms, how is it that they still possess them?*

A cue to answering this question is the current insight that self-repair does not necessarily rely on genetically-coded biochemical processes alone. After all, the bisected tail section of a worm also regenerates its severed head; the cellular differentiation that regenerates the amputated limb of the newt begins not at a point next to the wound, but at the opposite end. Processes of self-regeneration, it appears, proceed on the full set of information that was involved in building the organism, rather than relying on the partial set of genetic information that natural selection could have favored and the biochemistry of specialized cells would have transmitted.

Yet, if we assume that self-repair makes use of the full set of information coding the morphology of the organism, we face another puzzle: *How do organisms come by their full set of morphological information?*

IN SUMMARY . . .

The deepening puzzles facing classical Darwinism mark a turning point in the contemporary sciences of life. A number of front-line investigators are coming to the conclusion that biological evolution must involve extragenetic factors in addition to genetic ones.

The assumption is not new. In the last century German poet-scientist Wolfgang Goethe spoke of an *Urpflanze* after which all plants are patterned, and in the 1950s mathematician Hermann Weyl postulated immaterial factors in the nature of ideas, images or building plans. More recently biologist Alister Hardy speculated about a psychic blueprint that would be shared by all members of a species, Jean Dorst suspected a basic design in evolution, Gordon Rattray Taylor suggested that there must be a built-in tendency to self-assembly in the biological sphere, and paleontologist Roberto Fondi concluded that something like biological archetypes must guide evolution in the living world. However, as to the nature and origin of such an archetypal design, pattern, or information, biologists offer but timid guesses, or frankly speculative hypotheses.

CHAPTER 8

MYSTERIES IN THE MANIFESTATIONS OF MIND

THE PROBLEMS that face science and scientists in regard to the phenomena of mind are even deeper and more difficult than those that crop up in other spheres of observation and experience. Scientists cannot answer the fundamental question, why some of our cerebral functions should be accompanied by conscious experience. Nevertheless science, as we have seen, is not stymied by this. It proceeds on the assumption that brain and mind, if not necessarily identical, are at least closely linked. This means that, while philosophical queries such as "Why is it that we possess consciousness?" and "What is consciousness in and of itself?" elude scientists, they can track the more modest question, "With what kind of neural functions and mechanisms is consciousness associated?"

The latter query is now the topic of a major research effort; a frontal attack on the relationship between mind and brain. Brain scientists probe the deepest recesses of the neocortex with instruments such as microelectrodes, magnetic resonance imaging and positron emission tomography. These techniques can disclose many of the physiological mechanisms that are associated with the manifestations of consciousness in human beings.

The current wave of scientific consciousness research got underway in 1990, when Sir Francis Crick and his colleague Christoph Koch proclaimed that the time was ripe for a concentrated attempt to understand the phenomenon of consciousness. Consciousness, they said, is synonymous with awareness, and awareness always involves a combination of attention and short-term memory. Investigators should focus on visual awareness, as the visual system has already been well mapped in animals and in humans. If the neural mechanisms underlying visual awareness could be better understood, more complex and subtle mind-phenomena, such as the uniquely human feature of self-awareness (being conscious of being conscious) may be tracked as well.

These ideas have catalyzed a host of effort in consciousness research, as well as a great deal of controversy. It is not clear, some investigators noted, whether the kind of "electrophysiological" theory called for by Crick would suffice to explain consciousness. It may also be that studying the brain by itself is not enough: the entire body may be involved in every act of conscious mentation. If so, a neural model of consciousness would have to be completed by cognitive, and perhaps even by social, theories.

Physicists such as Roger Penrose and Henry Stapp have taken a different tack: they seek the key to understanding consciousness in the quantum processes that link electrons and other micro-particles within the networks of the brain. Though at first ignored and then attacked, the microphysical approach to consciousness has gained numerous adherents. The principal promise of the approach is to give an account of our sensation of free will. The brain, according to Penrose, exploits the nondeterministic effects of the quantum world (associated with the "collapse of the wave-function") to create processes that are free from the outset. This would account for our experience of making choices in light of our own will. An additional promise of the quantum approach to brain research is to explain how separate, and sometimes comparatively

distant, parts of the brain can be highly, and as it seems instantaneously, synchronized. The adherents of this approach claim that nonlocality—the apparent ability of a particle to be simultaneously in more than one location—may be characteristic of a number of processes in the brain as well.

Though the current assault on consciousness is promising, scientists are only scratching at the surface of the complex neurological processes that underlie human consciousness. The difficulties, as we have already remarked, are considerable. The gray matter we carry in our cranium is a highly integrated system topped by a cortex constituted as a six-layered sheet of some 10 billion neurons with up to a million billion connections. Regions of this super-complex brain are composed of groups and networks of neurons, where individual neurons are connected by synapses through their dendrites to the dendrites of other neurons.

The entire system works as a whole, yet it has a hierarchical structure. The most basic, innermost component of the brain's hierarchy is made up of the brain stem, the thalamus, and the hypothalamus. These relatively "hard wired" elements control the rudimentary bodily functions, the secretion of hormones, and automatic behavior processes. They constitute what Paul MacLean described as the "reptilian brain."

The middle component of the cerebral hierarchy is the "paleomammalian brain." It is already highly developed in some lower mammals, such as rodents. Its internal structure is dominated by the limbic system, where neurons are arranged in complex feedback loops. This system regulates the emotions, as well as the basic emotionally charged cognitions and behaviors. The highest element of the brain's hierarchy, called the "neomammalian brain," is present only in *Homo,* and in the primates, next to *Homo* the most highly evolved animals. It is dominated by the cerebral cortex. Here cells are arranged in columns that intersect the surface; thus the cortex can grow only if the surface area increases, allowing the placement of more columns. In our own species the surface of the cortex has grown into complex, highly convoluted folds. Beneath the columns there is a network of nerve fibers interconnecting the columns.

A typical feature of the human neocortex is its split into a right and a left hemisphere, connected by a massive bundle of fibers known as the corpus callosum. In healthy individuals the two hemispheres operate jointly, though they each have somewhat different functions. The left hemisphere operates in a sequential mode, connecting cause and effect and undergirding ordinary common sense and its linear processes of thought. It is the center of ordinary speech, a typically linear cerebral activity. The right hemisphere is keyed to correspondences and relationships; it processes complex data simultaneously. It charts the emotional nuances of perception and cognition, but has a limited syntax—it is oriented to images and not to words.

Findings such as these relate both higher and lower mental functions to processes in the brain. But, except in the case of relatively simple and basic elements of mentation, neuroscience's state of the art does not permit the creation of detailed models showing how processes in the brain would produce the corresponding mental functions: the cerebral details of the higher functions remain essentially unknown. Among the most mysterious of these functions are abstract thought, subtle emotional states, and memory. Memory of the long-term variety is perhaps the most mysterious of all.

THE MYSTERY OF LIFETIME MEMORY

Human beings, it appears, can store their experiences and impressions both temporarily and in the long run. Short-term memory is relatively well understood in reference to the formation and reformation of neuronal networks in the cortex, but long-term memory remains a puzzle. Yet the evidence for it is accumulating. In the past, exceptional feats of long-term memory surfaced through associations; for instance, when Swan, the hero of Marcel Proust's famous novels, recalled his childhood when drinking a familiar cup of tea. Almost anyone can suddenly recall episodes from the past, going back to the age of four or five, or perhaps even earlier. But clinical work by psychologists and psychotherapists has uncovered evidence that most people can recall events in their life that extend much further back. Many people show the trace of traumas or other unusual events that befell them up to the moment they were born; and the impact of physical and emotional stress on the mother during the period of gestation can show up in the psychological make-up of individuals throughout life.

There are now fresh, and seemingly more esoteric elements of long-term memory surfacing in the investigation of the so-called altered states of consciousness (ASCs). Of these, memory is a major component in both near-death experiences (NDEs) and in medically initiated and supervised regression analysis. These altered state experiences provide impressive testimony regarding the possibility of extremely long-term memory. The scope of such memory would be truly staggering. John von Neumann calculated that the amount of information an individual accumulates during his or her lifetime comes to about 2.8×10^{20} (that is, 280,000,000,000,000,000,000) "bits." *How could a 10-centimeter diameter brain hold that much information?*

Long-term memory is a major mystery; the evidence for it deserves a closer look.

Since Elisabeth Kübler-Ross' classic studies, NDEs have been systematically investigated by clinical psychologists and specialized researchers. It appears that people who come close to death undergo a remarkable experience that has a distinct memory component. Raymond Moody, Jr., who pioneered the systematic study of NDEs, concluded that it is now "clearly established" that the experience of a significant proportion of the people who are revived following close calls with death is quite similar from case to case,

regardless of the patient's age, sex, religious, cultural, educational
or socioeconomic background. The experience—which includes a
panoramic re-play of one's entire life—is more widespread than is
generally recognized: a 1982 survey conducted by George Gallup, Jr.,
found that some 8 million adults in the U.S. alone have undergone
them. Thirty-two percent of the people surveyed reported that "life-
reviews" were a part of their near-death experience.

British NDE researcher David Lorimer distinguished two kinds
of near-death recall: panoramic memory, and the life-review itself.
Panoramic memory, he claims, consists of a display of images and
memories with little or no direct emotional involvement on the part
of the subject; while life-review, although superficially similar,
involves emotional involvement and moral assessment as well. The
clarity of mental processes is noteworthy in both memory
processes. Recall is especially vivid in panoramic memory, where
there is a remarkable speed, reality and accuracy in the images that
flash across the mind. The time-sequence of the memories may
vary: some start in early childhood and move toward the present;
others start in the present and move backward to childhood. Still
others come superposed, as if in a holographic clump. To the subjects it appears that everything they have ever experienced in their
lifetime is being recalled; no thought, no incident, appears to have
been lost.[1]

Amazingly enough, our brain seems to have access to a store
of information that is even larger than that which we have accumulated in a lifetime. On this score the evidence is distinctly more
controversial, but it is by no means negligible. The most credible
strands are furnished by practicing psychotherapists. By "regress-
ing" patients to early childhood, the therapists often find that they
can proceed still further back in time, to experiences of the womb
and of birth. Sometimes they can go back further still, to events that
appear as if they stem from prior lifetimes. Some patients can recall
several past lives that together cover a vast time span. According to

Thorwald Dethlefsen, a famous if controversial therapist in Munich, Germany, the series of "reincarnations" may encompass hundreds of lifetimes and span 12,000 years. In the USA, Stanislav Grof, the famous Czech-born psychiatrist, hypno-regressed subjects down to the state of animal ancestors.

Patients of all ages tell stories of prior-life experiences, often associated with present problems and neuroses. Dethlefsen's case histories include the story of a patient who could not see in an otherwise functional eye; he came up with the memory of being a medieval soldier whose eye was pierced by an arrow. A patient of pioneer investigator Morris Netherton, suffering from ulcerative colitis, relived the sensations of an eight-year-old girl shot at a mass grave by Nazi soldiers. And New York therapist Roger Woolger's patient, who complained of a rigid neck and shoulders, recalled that as a Dutch painter he had committed suicide by hanging.

The images and experiences that surface from these mysterious sources often have a marked therapeutic effect: many psychic and some bodily ills seem to be the result of traumas that appear to have been experienced in previous lifetimes. To recall and relive such events releases "karmic bonds": feelings of guilt and anxiety that appear to have been carried over from earlier existences.

Past-life findings have been questioned. Investigators have uncovered evidence that in some cases subjects who recalled a particular image or event from a previous lifetime actually had prior information about the given persons, times, or places. In some cases, however, the information produced by regressed subjects contained elements that were not likely to have been available to them in their present existence. These remarkable data include obscure (but subsequently verified) historical and geographical particulars and the personal histories of people unknown to the regressed subjects, many of whom seem to have lived in distant lands and in times long gone. Moreover, most subjects not only remember but actually *relive* experiences in the regressed state, and their emotional tone and physiological responses are transformed beyond all reasonable bounds of accident or simulation. For example, a person regressed to early childhood can exhibit the sucking reflex and other, so-called axial, reflexes, and even the fan-like extension of the toes that occurs in infants when the lateral part of the sole is stimulated by a sharp object.

Ian Stevenson, a reputable medical doctor in the U.S., had as many as 2,000 children recount past-life experiences. He concluded that more children may have abundant memories of previous existences than we realize. In most cases we meet the likely subjects only later, by which time such memories have faded or entirely slipped away. Children who do refer to previous lifetimes do so between the ages of two and five—the average age of disclosure is 38 or 39 months. Before the age of two they lack the vocabulary and the verbal skills to communicate, and from the age of five heavy layers of verbal information cover the images in which the memories are conveyed. Past-life memories recede whether the parents encourage the children to remember them or forbid them to do so.

During the three-year window in time when communication is possible, children's past-life memories tend to cluster around the events of the last year, month, or day in the life of the person with whom they identify. Sometimes the past-life memories appear more real than the present-life experiences. Stevenson reported that among the very first words of a Turkish child were "What am I doing here? I was at the port." When he could say more, the child described details of the life of a dockworker who was killed in an accident while asleep in the hold of a ship. Nearly three-quarters of the subjects claim to remember how the person of the previous life died, and remember it more often when the death was violent than when it occurred by natural causes. The question must be faced: *where do these memories come from?*

THE MYSTERY OF TRANSPERSONAL COMMUNICATION

There is yet another puzzling facet of experience: this is "transpersonal" contact and communication, not just by young children or sensitives, but by almost any individual.

Conservative investigators tended to insist that people can communicate only through gestures, facial expressions, and by means of language, that is, in the "standard mode." There is evidence, however, that communication can take place also in decidedly non-standard modes. Insofar as these involve the sending and receiving of messages beyond the range of eye and ear and other sensory organs, they come under the heading of transpersonal communication. Such communication seems to involve some form of extrasensory perception (known as ESP).

Telepathy, the most common form of ESP, may have been widespread in so-called primitive cultures. It appears that in many tribal societies shamans were able to communicate telepathically, using a variety of techniques to enter the altered states of consciousness that seem required for it, including solitude, concentration, fasting, as well as chanting, dancing, drumming, and the use of psychedelic herbs. Not only shamans, but entire tribes seem to have possessed the gift of telepathy. To this day, many Australian aborigines appear to be informed of the fate of family and friends, even when out of sensory communication range with them. Anthropologist A.P. Elkin noted that a man, far from his homeland, "will suddenly announce one day that his father is dead, that his wife has given birth to a child, or that there is some trouble in his country. He is so sure of his facts that he would return at once if he could."

Aside from anthropological data, largely anecdotal and unrepeatable, scientific evidence for various kinds of transpersonal contact and communication comes from laboratory research based on controlled experiments.

The scientific investigation of ESP dates back to J.B. Rhine's pioneering card- and dice-guessing experiments at Duke University in the 1930s. Recently experiments have become more sophisticated, and experimental controls more rigorous; physicists have often joined psychologists in designing the tests. Explanations in terms of hidden sensory cues, machine bias, cheating by subjects, and experimenter error or incompetence have all been considered, but they were found unable to account for a number of statistically significant results.

In the 1970s two physicists, Russell Targ and Harold Puthoff of the Stanford Research Institute, carried out some of the best-known experiments on thought and image transference. They wished to ascertain the reality of telepathic transmission between different individuals, one of whom would act as "sender" and the other as "receiver." The scientists would place the receiver in a sealed, opaque and electrically shielded chamber, and the sender in another room where he or she was subjected to bright flashes of light at regular intervals. Electroencephalograph (EEG) machines would register the brain-wave patterns of both. As expected, the sender exhibited the rhythmic brain waves that normally accompany exposure to bright flashes of light. But, after a brief interval the receiver also began to produce the same patterns, although he or she was not exposed to the flashes and was not receiving sense-perceivable signals from the sender.

A particularly striking example of this kind of communication is the work of Jacobo Grinberg-Zylberbaum at the National University of Mexico. In more than 50 experiments performed over the past five years, Grinberg-Zylberbaum paired his subjects inside sound- and electromagnetic radiation-proof "Faraday cages." He asked them to meditate together for 20 minutes. Then he placed the subjects in separate Faraday cages where one of them was stimulated and the other not. The stimulated subject received stimuli at random intervals in such a way that neither he nor she, nor the experimenter, knew when they were applied. The non-stimulated subject remained relaxed, with eyes closed, instructed to feel the presence of the partner without knowing anything about his or her stimulation.

In general, a series of 100 stimuli was applied—flashes of light, sounds, or short, intense but not painful electric shocks to the index and ring fingers of the right hand. The EEG of both subjects was then synchronized and examined for "normal" potentials evoked in the stimulated subject and "transferred" potentials in the non-stimulated subject. Transferred potentials were not found in control situations where there was either no stimulated subject or when a screen prevented the stimulated subject from perceiving the stimuli (such as light flashes); or else when the paired subjects did not previously interact. However, in experimental situations with stimulated subjects and with interaction, the transferred potentials appeared consistently in some 25 percent of the cases. A particularly poignant example was furnished by a young couple, deeply in love. Their EEG patterns remained closely synchronized throughout the experiment, testifying to their report of feeling a deep oneness.

In a limited way, Grinberg-Zylberbaum could also replicate his results. When a subject exhibited the transferred potentials in one experiment, he or she usually exhibited them in subsequent experiments as well.

Grinberg-Zylberbaum's experiment is not unique: in the past several years it has been matched by hundreds of similar experiments. They provide significant evidence that identifiable and consistent electrical signals occur in the brain of one person when a second person, especially if he or she is closely related or emotionally linked, is meditating, provided with sensory stimulation, or attempts to communicate with the subject intentionally.

Transpersonal experiences also occur outside the laboratory; they are particularly frequent among identical twins. In many cases one twin feels the pain suffered by the other, and is aware of traumas and crises even if he or she is on the other side of the world. Besides "twin pain," the sensitivity of mothers and lovers is equally noteworthy: there are countless stories of mothers having known when their son or daughter was in grave danger, or was actually involved in an accident.

Transpersonal contact is not limited to twins, mothers and lovers—the kind of closeness that a therapeutic relationship creates between therapist and patient seems to suffice already. A number of psychotherapists have noted that, during a session, they experience memories, feelings, attitudes, and associations that are outside the normal scope of their experience and personality. At the time these strange items are experienced they are indistinguishable from the memories, feelings and related sentiments of the therapists themselves; it is only later, on reflection, that they come to realize that the anomalous items stem not from their own life and experience, but from their patient.

It appears that in the course of the therapeutic relationship some aspect of the patient's psyche is projected into the mind of the therapist. In that location, at least for a limited time, it integrates with the therapist's own psyche and produces an awareness of some of the patient's memories, feelings, and associations. (The inverse can also take place: patients can obtain undisclosed details of the life and personality of their therapist.) Known as "projective identification," the patient-to-therapist transference can be useful in the context of analysis: it can permit the patient to view what was previously a painful element in his or her personal consciousness more objectively, as if it belonged to somebody else.

The psychotherapeutic experience, jointly with the experience of twins and lovers and of a wide variety of subjects in controlled experiments, raises another intriguing question. *Could it be that most people—and not just specially gifted sensitives—have the ability to "enter" into the brain and mind of another person, especially if they are related or emotionally close?*

The transference of feelings and associated memories and attitudes is not the only kind of transpersonal contact and communication for which there is significant evidence. Another variety involves the transmission of images.

In addition to thought-transference experiments, Targ and Puthoff have also conducted so-called "remote-viewing" tests. In these experiments sender and receiver are separated by distances that preclude any form of sensory communication between them. At a site chosen at random, the sender acts as a "beacon"; the receiver then tries to pick up what the beacon sees. To document his or her impressions, the receiver gives verbal descriptions, at times accompanied by sketches. In Targ's and Puthoff's experiments independent judges found that the descriptions of the sketches matched on average 66 percent of the time the characteristics of the site that was actually seen by the beacon.

Remote-viewing experiments reported from other laboratories involved distances from half a mile to several thousand miles. Regardless of where they were carried out, and by whom, the success rate was generally around 50 percent—considerably above random probability. The most successful viewers appeared to be those who were relaxed, attentive, and meditative. They reported that they received a preliminary impression as a gentle and fleeting form which gradually evolved into an integrated image. They experienced the image as a surprise, both because it was clear and because it was clearly elsewhere.

Images can also be transmitted while the receiver is asleep. Over a full decade, Stanley Krippner and his associates carried out "dream ESP experiments" at the Dream Laboratory of Maimondes Hospital in New York City. The experiments followed a simple yet effective protocol. The volunteer, who would spend the night at the laboratory, would meet the sender and the experimenters on arrival and have the procedure explained to him or her. Electrodes were then attached to the volunteer's head to monitor brain waves and eye movements; there was no further sensory contact with the sender until the next morning. One of the experimenters would throw dice that, in combination with a random number table, gave a number that corresponded to a sealed envelope containing an art print. The envelope was opened when the sender reached his or her private room in a distant part of the hospital. The sender then spent the night concentrating on the print.

The experimenters woke the volunteers by intercom when the monitor showed the end of a period of rapid-eye-movement (REM) sleep. The subject was then asked to describe any dream he or she might have had before awakening. The comments were recorded, as were the contents of an interview the next morning when the subject was asked to associate with the remembered dreams. The interview was conducted double blind—neither the subject nor the experimenters knew which art print had been selected the night before.

Using data taken from the first night that each volunteer spent at the dream laboratory, the series of experiments between 1964 and 1969 produced 62 nights of data for analysis. They exhibited a significant correlation between the art print selected for a given night and the recipient's dreams on that night. The score was considerably higher on nights when there were few or no electrical storms in the area and sunspot activity was at a low ebb—that is, when the Earth's geomagnetic field was relatively undisturbed. *Is it conceivable that under suitable conditions one person can send images directly into the mind of another?*

A different type of experiment investigated the degree of harmonization between the left and right hemispheres of the subject's neocortex. In ordinary waking consciousness the two hemispheres—our language-oriented, linear-thinking, rational "left brain" and our gestalt-perceiving, intuitive "right brain"—exhibit uncoordinated, randomly diverging wave patterns in the EEG. When the subject enters a meditative state of consciousness, these patterns become synchronized, and in deep meditation the two hemispheres fall into a nearly identical pattern. Even more remarkably, in deep meditation not only the left and right brains of one and the same subject, but also the left and right brains of *different* subjects manifest identical patterns. Experiments in Italy with up to

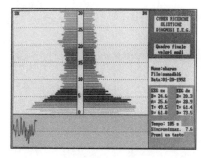

FIGURE 5A Printout of the EEG waves of an average person in the ordinary state of consciousness. The left and right brain hemispheres show insignificant correlation (synchronization value 7.6 percent) and no specifically harmonic patterns. (The printout encompasses theta, alpha, beta, and delta waves, given in terms of the known frequency regions from 0 to 30 waves per second.)

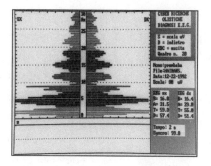

FIGURE 5B Printout of a practiced meditator in a state of deep meditation. The left and right hemispheric EEG patterns have a pronounced harmonic element and they are highly synchronized (synchronization value 99.8 percent).

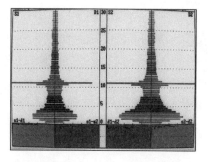

FIGURE 5C The EEG patterns of two subjects meditating together but without sensory contact with each other. The left and right brain hemispheres of each subject (left and right side of the image) are quasi-identical in both subjects (cross-person synchronization value over 90 percent).

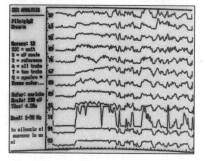

FIGURE 5D EEG of 12 almost completely synchronized people in deep meditation showing a sense of unity. The average synchronization is 81.2 percent.

12 subjects simultaneously displayed an astonishing synchroniza-tion of the brain waves of the entire group.[2]

There is no known limit to the size of group that could thus be brain-synchronized. Nitamo Montecucco, an Italian experimenter who worked extensively in India, speaks of vast "Buddha-fields" resulting from the simultaneous meditation of a large number of people. *Could it be that as well as single individuals being able to spontaneously affect the brain and mind of another, many people meditating together could develop some kind of collective consciousness?*

Medicine has been the scene of a related form of transpersonal contact and communication. It is known as distance diagnosis. Sensitive diagnosticians are given a few basic particulars of their patient: name and birth date are often sufficient. Even without medical training, they are able to arrive at a surprisingly accurate diagnosis of what is wrong with the patient.[3]

Distance diagnosis is now fairly common. In the U.S., neuro-surgeon Norman Shealy provided impressive evidence of it in his book *The Creation of Health*. He would telephone the name and birth date of a patient sitting in his office in Missouri to clairvoyant diagnostician Carolyn Myss, located in distant New Hampshire. Thereupon she would provide him with the diagnosis. Dr. Shealy claims that in the first 100 cases her correct diagnosis rate was 93 percent. *Is it also possible, then, that one person can "see into" an-other across vast distances, and tell what is wrong with him or her?*

A further variety of transpersonal communication involves the transmission of actual bodily effects from one individual to another. Transmissions of this kind came to be known as "telesomatic": they consist of physiological changes that are triggered in the targeted person by the mental processes of another. Here, too, intervening distance seems to make little or no difference.

Traditionally, telesomatic effects were produced by specially gifted natural healers, who would "send" what they claimed to be subtle forms of energy to their patients. Being largely anecdotal, such effects were dismissed by the medical community. Lately, however, they have been noted in laboratory experiments where a large number of trials and test subjects allows a reliable quantitative analysis of the results. William Braud and Marilyn Schlitz of the Mind Science Foundation in San Antonio, Texas, have carried out hundreds of telesomatic experiments with rigorous controls: they tested the impact of the mental imagery of senders on the physiology of receivers. The latter were both distant and unaware that such imagery was being directed to them.

Braud and Schlitz claim to have established that the mental images of a person can "reach out" over space and cause changes in the physiology of a distant person—effects comparable to those one's own mental processes produce in one's own body. Their experiments show that people who attempt to influence their own bodily functions are only slightly more effective than people who attempt to influence other people's physiology from a distance. Over several cases involving a large number of individuals the difference between remote influence and self-influence was insignificant: remote telesomatic influence by a second person proved to be nearly as effective as psychosomatic self-influence by the same person.

It is curious that telesomatic effects can be transmitted also in the form of what anthropologists call "sympathetic magic." Shamans, witch doctors and others who practice such magic—voodoo is a familiar example—act not on the person they target, but on an effigy of that person, such as a doll. The practice is widespread among traditional people; the rituals of American Indians made use of it as well. In his famous study, *The Golden Bough,* Sir James Frazer noted that practices among Native Americans included drawing the figure of a person in sand, ashes, or clay, and then pricking it with a sharp stick or doing it some other injury. The corresponding injury was said to be inflicted on the person that figure represented. Experimental parapsychologists Dean Radin and colleagues at the University of Nevada decided to test this effect under controlled laboratory conditions.

In the experiments the subjects would create a small doll in their own image, and along with the doll would also include various small objects, pictures, jewelry, an autobiography, and personally meaningful tokens that would "represent" them. They would also provide a list of what made them feel nurtured, calm and comfortable. This information was used by the active experimenter (called the "healer," for the effects tested were beneficial rather than malign) to create a sympathetic connection to the "patient." The latter was wired up so as to monitor the activity of his or her autonomous nervous system—electrodermal activity, heart rate, and blood pulse volume—while the healer was in an acoustically and electromagnetically shielded room in an adjacent building. The healer would place the doll and the other small objects on the table in front of him and concentrate on them while sending—in a random sequence—"nurturing" (active healing) and "rest" messages to the subject.

A typical experimental session consisted of 5 nurturing and 5 rest periods of 60 seconds each, followed by an 11 second interval. It turned out that the electrodermal activity of the patients together with their heart rate were significantly different during the active nurturing periods than during the rest periods, while blood pulse volume was significant for a few seconds in the middle of the 60 second nurturing period. Heart-rate and blood flow indicated a "relaxation response"—which made sense since the healer was attempting to "nurture" the subject via the doll. In turn, the higher rate of electrodermal activity showed that the patients' autonomic nervous system was becoming aroused. This otherwise puzzling result was cleared up when it was noted that the healers nurtured the patients by rubbing the shoulders of the dolls that represented them, or stroked their hair and face. This, it appears, had the effect of a "remote massage" for the patients.

Radin and his colleagues concluded that the local actions and thoughts of the healer were mimicked in the remote patient almost as if healer and patient were next to each other. This confirms the finding that telesomatic effects work much the same as psychosomatic ones, notwithstanding the distance. And it appears that they work much the same even if the intention directed at the subject is mediated by a doll or another object that would "represent" him.

Spontaneous (as contrasted with intentional) telesomatic effects have been noted in regard to entire groups of people. According to a traditional Hindu notion, when a significant number of people meditate in a community, the non-meditators are affected also. In 1974 the Maharishi Mahesh Yogi took up this idea. He suggested that if but 1 percent of a population were to meditate regularly, the effects would be felt also on the remaining 99 percent. Empirical studies, by Garland Landrith and David Orme-Johnson among several others, showed that the "Maharishi effect" is statistically significant. There appears to be more than random correlation between the number of meditators in a community and community crime rates, incidence of traffic fatalities, deaths due to alcoholism, and even levels of pollution.

Cardiologist Randolph Byrd, a former professor at the University of California, carried out an intentional variant of the telesomatic group effect, using prayer rather than meditation. His ten-month computer-assisted study concerned the medical histories of patients admitted to the coronary care unit at San Francisco General Hospital. Byrd formed a group of experimenters made up of ordinary people whose only common characteristic was a habit of regular prayer in Catholic or Protestant congregations around the country. The selected people were asked to pray for the recovery of a group of 192 patients. Another set of 210 patients, for whom nobody prayed in the experiment, made up the control group. Rigid criteria were used: the selection was randomized and the experiment was carried out double blind, with neither the patients nor the nurses and doctors knowing which patients belonged to which group.

The experimenters were given the names of the patients, some information about their heart condition, and were asked to pray for them every day. They were not told anything further. Since each experimenter could pray for several patients, each patient had between 5 and 7 people praying for him or her. The results were statistically significant. The prayed-for group, it turned out, was 5 times less likely than the control group to require antibiotics (3 compared to 16 patients); it was three times less likely to develop pulmonary edema (6 versus 18 patients); none in the prayed-for group required endotracheal incubation (while 12 patients in the control group did), and fewer prayed-for patients died (though this particular result was statistically not significant). It did not matter how close or far the patients were to those who prayed for them, nor did the manner of praying make any difference. Only the fact of concentrated and repeated prayer seems to have counted, regardless of who the prayer was addressed to and where it took place.

Literally hundreds of experiments of this kind have been carried out by now. They raise another intriguing possibility. *Could the focused collective consciousness of a group of people affect the bodily condition of other people—even a large number of other people?*

THE MYSTERY OF SPONTANEOUS CULTURE LINKS

Evidence of spontaneous culture links comes from history. It appears that from time to time significantly similar achievements have been produced by entire cultures even though they were not in ordinary communication with each other, and may not even have known of each other's existence.

To begin with, there have been cultures in widely different locations that have developed an impressive array of highly similar tools. The Acheulian hand-axe, for example, was a widespread tool of the Stone Age: it had a typical almond or tear-shaped design chipped into symmetry on both sides. In Europe the axe was made of flint, in the Middle East of chert, and in Africa of quartzite, shale, or diabase. Its basic form was functional, yet the agreement in the details of its execution in virtually all traditional cultures cannot be readily explained by the coincidental discovery of utilitarian solutions to shared needs—trial and error is unlikely to have produced such similarity of detail in so many and so far-flung populations.

Many artifacts seem to have leaped across space beyond the range of direct culture contact. Giant pyramids were built in ancient Egypt and in pre-Columbian America with remarkable agreement in design. Crafts, such as pottery-making, have taken much the same form in all cultures. Even the technique of making fire brought forth implements of the same basic design in different parts of the world. Ignazio Masulli, a reputable historian at the University of Bologna, has made an in-depth study of the pots, urns, and other artifacts produced by indigenous cultures dating from the 5th and 6th centuries BC in Egypt, Persia, India, and China. Masulli found that there is no reasonable explanation for the striking recurrence of their basic design: direct contact between these cultures is dismissed by archeological research, and functional utility would have allowed for a far wider range of solutions than those that were actually adopted. The phenomenon is widespread. While each culture added its own embellishments, Aztecs and Etruscans, Zulus and Malays, classical Indians and ancient Chinese all fashioned their tools and built their monuments as if following a shared pattern or archetype.

More than physical artifacts, entire culture patterns have emerged more or less simultaneously, yet independently of each other. The great breakthroughs of classical Hebrew, Greek, Chinese, and Indian culture occurred in widely scattered regions, yet they occurred practically simultaneously. The major Hebrew prophets flourished in Palestine between 750 and 500 BC; in India the early *Upanishads* were composed between 660 and 550 BC and Siddhartha the Buddha lived from 563 to 487 BC; Confucius taught in China around 551–479 BC; and Socrates lived in Hellenic Greece

from 469 to 399 BC. Just when the Hellenic philosophers created
the basis of Western civilization in Platonic and Aristotelian philosophy, the Chinese philosophers founded the ideational basis of
oriental civilization in the Confucian, Taoist and Legalist doctrines.
While in the Hellas of the post-Peloponnesian Wars period, Plato
founded his Academy and Aristotle his Lyceum, and scores of itin-
erant sophists preached to and advised kings, tyrants and citizens,
in China the similarly restless and inventive "Shih" founded schools,
lectured to crowds, established doctrines, and maneuvered among
the scheming princes of the late Warring States Period.

Simultaneous cultural achievements are not limited to classical
civilizations: they also occur in modern times. Even within the dis-
ciplined domain of science, there are documented cases of insights
occurring at the same time to different investigators who did not
know of each other's work. Among the most celebrated of these
cases are the simultaneous and independent discovery of the
calculus by both Newton and Leibniz, the likewise simultaneous
and independent elaboration of the fundamental mechanisms of
biological evolution by Darwin and Wallace, and the concurrent
invention of the telephone by Bell and Gray.

There have been instances where insight and discovery have
leapt across different branches of the same culture. Just at the time
when Newton was using a prism to break down the shafts of light
that entered the windows of his Cambridge lodgings, Vermeer and
other Flemish artists were exploring the nature of light entering
through colored window and door-panes. While Maxwell was for-
mulating his electromagnetic theory, according to which light is
produced by the reciprocal revolution of electrical and magnetic
waves, Turner was painting light as swirling vortices. In recent
years physicists have been exploring many-dimensional spaces in
supersymmetry theories—and simultaneously, and apparently
entirely independently, avant-garde artists began to experiment
with visual superposition on their canvases, representing as many
as seven spatial dimensions.

Space and time, light and gravity, mass and energy have all been explored by physicists and artists, sometimes at the same time, sometimes one preceding the other, but seldom if ever in conscious knowledge of each other. In his *Art and Physics: Parallel Visions in Space, Time, and Light,* Leonard Shlain provided numerous illustrations of the power of artists to mirror, and frequently to anticipate, the conceptual breakthroughs occurring in the minds of physicists, without themselves knowing anything about physics and the concerns of its investigators. *Can all these parallelisms be dismissed as mere coincidence?*

IN SUMMARY . . .

Confounding the classical tenet, that everything knowable about the mind can be ultimately referred to sensory experience, scientists are now confronted with data from a variety of fields that speak to the reality of essentially transpersonal modes of contact and communication. It may be that the human mind is more widely "informed" than has been generally recognized. Our sources of information are not limited to the bodily sense organs: there may be items entering the mind that exceed the range of ordinary perception. Stanislav Grof suggested that we should complete the standard cartography of the human mind with additional elements— those that hitherto had been the province of mystical and esoteric disciplines. Both a "perinatal" and a "transpersonal" domain need to be added to the standard "biographic-recollective" domain of the psyche. The transpersonal domain, according to Grof, can mediate connection between our mind and practically any part or aspect of the phenomenal world.

Grof's suggestion needs to be taken seriously. It calls for a revision of our standard concept of the mind, but not for a plunge into mysticism and metaphysics. An enlarged cartography does not mean that the mind is an immaterial entity, unrelated to the brain. The indicated assumption is merely that our brain is sensitive to information beyond the range of our body's sensory organs. The new consciousness researchers realize this already, though conservative scientists may find it hard to accept. Yet science is an open enterprise, and if one set of investigators does not rise to the challenge, then another does. The result promises to be a new appreciation of the mind as a powerful organ. Making full use of it we shall be able to, as physicist William Tiller suggests, "open the roof to the sky," rather than remain constrained to view the world through "five slits in the tower."

NOTES

1. NDEs are similar to the experience of subjects under hypnosis in that they demonstrate the possibility of a quasi-total recall of a person's prior experiences. But NDEs are less controversial than recall under hypnosis: in that state the subject is open to conscious or unconscious suggestion by the hypnotist, and that could taint the evidence. This problem does not surface in NDEs.

2. This was witnessed by the writer himself. A personal computer was hooked up with the EEG and a specially designed program analyzed the level of synchronization of the brain's two hemispheres. Tests with this "brain holotester" showed that when two subjects meditate simultaneously, the same synchronization effect is produced not only between their individual left and right hemispheres, but also *between* their respective brains. In deeply meditating subjects a quasi-identical fourfold synchronization emerged (left- and right-hemisphere synchronization within, as well as between, the subjects), even though they themselves did not see, hear, or otherwise sense each other.

3. Distance diagnosis has also been experienced by the present writer firsthand. In 1993 he was contacted by a group of accredited medical doctors in England who belong to the Psionic Medical Society. Their method of healing, called "psionic medicine," makes use of a sophisticated form of dowsing for diagnosis, and homeopathic remedies for treatment. But the relevant feature of their method is neither the one nor the other. It is the fact that diagnosis is not based on an analysis of the biochemical properties of the patient's organism, but on a field the psionic

practitioners take to be associated with the patient's organism. The "psi field" (also called "vitality field") is said to enter individuals at or near conception and to endure until death. Throughout the individual's lifetime it provides the cells in his or her body with the information they need to form tissues relevant to their location: the organism grows into the shapes and structures that are defined in the psi field. That field is not a self-enclosed system: it is sensitive to inputs from the individual's environment and also from his or her past. It carries features inherited from one's progenitors over several generations. Some weaknesses in the field (termed "miasmas" in standard homeopathic terminology) seem to be due not to illnesses contracted by the patient, but to maladies suffered by parents or grandparents.

Psionic treatment proceeds by acting on the patient's psi field, rather than directly on his or her body. This is borne out by the way diagnosis is carried out. It is not effected on the patient directly, but on a so-called "witness," which can be any sample of the patient's organism such as a strand of hair or a drop of blood. The witness can be analyzed at any time, and at any distance from the patient. The information it produces is not limited to the state of health of the patient at the time the sample was taken, but reflects his or her state of health at the time of the diagnosis: the sample gives ongoing information about the patient's ever-changing condition. This, obviously, would not be the case if only the unchanging (and indeed progressively degenerating) cellular or molecular structure of the sample were analyzed.

PART THREE

IN QUEST OF A NEW UNDERSTANDING

Far from there being an "end" of science,
our period will see the birth of a new vision,
a new science whose cornerstone encloses
the arrow of time; a science that makes
us and our creativity the expression of a
fundamental trend in the universe.

Ilya Prigogine, *World Futures* (1994)

CHAPTER 9

THE SEARCH FOR UNIFIED THEORIES:
1. IN THE NEW PHYSICS

CONTINUING OUR JOURNEY

SCIENCE'S established vision is blurring. Though the natural sciences of our day are more accomplished than ever, they are far from having solved all the mysteries and understood all there is to understand of the world. On the contrary, the confident image of the mid-20th century is fading: entire areas are missing, as if someone had removed pieces of a jigsaw puzzle and placed a milky sheet over the rest. Within the mainstream of the major disciplines scientists have been retreating into the safety of technicalities, leaving behind truly fundamental questions. As philosopher of science Carl Friedrich von Weizsäcker put it, "It is characteristic of physics as practiced nowadays, not to really ask what matter is, for biology not to really ask what life is, and for psychology not to really ask what soul is . . ."

This is now changing. Leading scientists have begun to re-examine their assumptions about the meanings that hide behind their observations and the equations by which they are described. When they encounter recurrent anomalies and paradoxes they no longer patch up the established theories, but look beyond them to bold new concepts and hypotheses.

This is much like what happened in the 16th century when the geocentric theory of the heavens was abandoned in favor of the heliocentric theory. The time-honored view, that the Sun and planets describe perfect circles around the Earth by being attached to revolving crystal spheres, could only be squared with observations by assuming that there are spheres within the spheres, cycles within the cycles. In the end the "epicycles" became so numerous, and their computation so complex, that the patched up theories had been discredited—the geocentric concept itself was finally abandoned. Copernicus, who firmly believed that nature loves simplicity, came out with the heliocentric theory, and this simpler hypothesis, though revolutionary, was adopted by the community of astronomers. The revolution created by Einstein at the turn of the 20th century was due to similar factors: the interpretation of physical phenomena in terms of Newton's classical mechanics became so cumbersome that the order and simplicity introduced by Einstein's relativity equations, abstract as they were, were greeted with almost audible sighs of relief.

A comparable process unfolds today in a number of scientific fields. No theory seems immune to disconfirmation: we have seen that in the course of this century anomalies have been encountered even in quantum physics and cosmology. As a result a new generation of physicists is actively seeking fresh approaches, exploring new concepts. Anomalies multiply also in the domain of biology, creating increasing pressure not only on the classical Darwinian theory, but also on its neo-Darwinian versions. The domain of mind, of course, has never been fully understood by science, but there have been scientists at times who believed that they had the basic notions well in hand, with a full understanding just down the road. The hold of such presumptions has now been loosened: paradoxical findings have come to light regarding the farther reaches of human mind and experience, findings that cannot be simply dismissed as illusory or extrascientific.

Gone now is the assurance that the fundamental features of the natural universe have already been discovered; the complacency typical of the late 19th century has almost vanished at the end of the 20th. More and more societies and associations are being created for the exploration of the anomalies encountered in scientific observation, and they are moving from the fringes of the science establishment toward its centers. Science is in the throes of another "revolution."

The scientific revolution now getting into gear is faster than the Copernican revolution and broader than the one initiated by Einstein. Its typical feature is the integration of a wide range of findings within a highly unified, simple if abstract theoretical framework. This is because in science one does not solve a puzzle and answer a question simply by tacking on a new proviso to an already established concept. At the critical juncture, when anomalies accumulate beyond the level where leading scientists can tolerate them, there is a leap to a new basic assumption—to a new "paradigm." Such a paradigm shift integrates the known as well as the anomalous findings of a given field in a theoretical framework at a deeper (or, if you prefer, higher) level.

Just as the theoretical framework advanced by Newton and Copernicus opened the way to the integration of the celestial with the earthly sphere, and as the basic concepts of 20th-century non-equilibrium thermodynamics paved the way toward the integration of physical and physicochemical systems with biological and even psychological and sociocultural systems, so the current revolution penetrates to a new level of theory construction. It seeks an understanding that is more coherent and integrated than that which underlies the fragmented approaches and languages of the puzzle-ridden classical disciplines.

In the history of science—from Galileo, Newton, Copernicus, and Kepler, to Einstein, Bohr, Jung, Guth, Hawking, and Pribram—significant progress has always involved both deeper and broader insights into the nature of empirical reality. The upcoming revolution sets forth the series initiated in the past: it lowers the floor of scientific inquiry, and extends its base.

The floor of scientific inquiry has been consistently lowered throughout the past several centuries. First the indivisible atom of Democritus was rediscovered by Dalton and Lavoisier as the basic constituent of gaseous matter. Then, when Dalton's atom proved fissionable, the floor was lowered to the atom of Rutherford, which consisted of a tinier nucleus and its orbital electrons. A still deeper basement has been reached this century at the level of Planck's constant, with the discovery of quarks, strings and the 200-odd elementary particles that came to light in high-energy experiments. And the field in which these progressively more minute and abstract entities are embedded—the so-called "zero-point field," of which more will be said below—has transformed from the passive Euclidean space of classical mechanics to the turbulent, potential-energy-filled quantum vacuum.

Where do we go from here? The new scientific revolution is still in its early phases, a fully-fledged theory is but a hopeful prospect. But there are indications that it is a realistic prospect, and that it projects a powerful new vision of the world. It is to these indications that our continuing journey takes us next.

THE "GUTs" OF THE NEW PHYSICS

The clearest indication for the coming paradigm shift is the intense search, under way in various disciplines, for a more integrated theory. Such a theory goes by many names: systemic, holistic, integrative, or simply "general." The term many scientists prefer is "unified." The uncontested paradigm for it is the field of grand unified theories (GUTs) in the new physics.

The objective of unification—though not "grand unification"—is familiar in the history of the physical sciences. In its time each major theory had unified the principal facts then known to the physics community. This was the case in the mechanics outlined by Galileo and the universal formulation of the Galilean theory at the hands of Newton; it was also the case in the electrodynamics of Maxwell and the thermodynamics of Boltzmann. In the beginning of this century Einstein contributed the crucial breakthrough that unified the suddenly anomalous world picture of 19th-century physics. This was the merit of Special Relativity, where the puzzles raised by classical physics found consistent and elegant resolution, and even more of General Relativity, where geometry and mechanics became fully and unexpectedly integrated. Space and matter, with the geometry of the one and the mechanics of the other, achieved a new and integral unity. The formerly mechanical force of gravity became an element of geometry; it was viewed as the effect of the curvature of space. The geometry of the latter, in turn, was traced to the distribution of matter. While it continued to be useful at times to think of space and matter as distinct entities, physicists became convinced that these form an inseparably integrated whole.

Not contented with the unification of geometry and mechanics, Einstein sought the further step that would integrate all known particles of matter with all known forces of space-time within the in itself timeless matrix of a unified field theory. But Einstein's attempt embraced only two of the four universal forces of interaction —gravitation and electromagnetism—and left the weak and strong nuclear forces out of account. That it ultimately failed was due to this mistaken assumption about the universal forces of nature, and not to the intrinsic unfeasibility of the enterprise. In creating grand unified theories physicists now encompass all four universal forces, together with the wide array of particles that came to light in the latter half of this century.

Grand unified theories offer a conceptual framework for unifying the particles that come to light in current experiments, as well as the forces that govern the interaction of the particles. Unification has become ever more necessary, seeing that the "elementary" particles proliferated extensively, and proved to be anything but elementary.

The Unification of Particles

In the 1920s only three elementary particles were known: the photon, the electron, and the proton. English physicist Ernest Rutherford then suggested that a further particle must be present in the nucleus: the neutron. When the existence of this particle was confirmed by experiment, the repertory of elementary particles had begun to expand. In 1930, in an attempt to explain the puzzling results of experiments on the decay of radioactive nuclei, the existence of the neutrino was suggested by Swiss quantum physicist Wolfgang Pauli. Twenty-five years later, the neutrino also was experimentally confirmed.

At that time quantum theory had already offered a good understanding of the outer shell of atoms, but the stability of the atomic nucleus remained puzzling. Japanese nuclear scientist Hideki Yukawa suggested that a new elementary particle was involved. Since its mass was predicted to lie between that of the proton and the electron, he named it meson. According to Yukawa's theory, the stability of the atomic nucleus is due to the constant exchange of mesons between protons and neutrons.

When experiments were designed to trace the meson, physicists discovered not one particle but a whole family that included muons and pions. And, as more powerful particle accelerators came on line, and nuclear collisions were investigated in cosmic rays high above the atmosphere, several further elementary particles were discovered. Some were found as experimentalists followed up predictions by theoreticians; others emerged unforeseen in the experiments.

The first elementary particles—the electron, the proton, the neutron, and the early mesons—appeared as expected and fitted into the then current theories of the structure of the atom. But as physicists moved the experiments to higher energy levels, observations refused to match theory. One mismatch involved the life-expectancy of exchange particles. Theory dictated that such particles should endure for only 10^{-23} seconds—a period during which a ray of light would have barely enough time to travel the width of an elementary particle—but experiment showed that the particles exist for 10^{-10} seconds: long enough for light to flash across an ordinary object. Since the particles endure 10 trillion times longer than expected, and are always produced in pairs, they came to be known as strange particles.

In order to create order among the many—strange and less strange—denizens of the emerging "particle zoo," Yale physicist Murray Gell-Mann suggested grouping particles in a particular eightfold way (he intended a reference to the eightfold path of the Buddha). This ordering was based on the theory that particles are made up of a more fundamental entity called the quark. Originally, there were thought to be three varieties of quark: the up, the down, and—there, too—the "strange." The proton, for example, consists of two up and one down quarks, the neutron of two down and one up quarks, and the exchange particles have a strange quark in addition. However, when more particles came to light, three quarks no longer sufficed and the family of quarks grew from three to six members. The sixth and last member of the family—the "top" quark—was discovered in high-energy experiments at Chicago's Fermilabs in the beginning of March, 1995.

Gell-Mann's theory of quarks—for which he received the Nobel Prize—resolved a persistent problem in the grouping of particles: while leptons (low-mass particles such as electrons and neutrinos) had a coherent symmetry group, hadrons (heavy particles such as protons and neutrons) did not. If, however, each hadron is composed of three quarks, also the hadron family can be integrated in symmetry groups in reference to combinations of quarks.

The Unification of Forces

Ordering the vast array of particles into coherent symmetry groups was a major achievement of mathematical physics, but genuine unification required that also the forces represented by the particles be unified. This endeavor was pioneered by Einstein in his unified field theory. His theory, even though it considered only the gravitational and the electromagnetic forces and was hence condemned to fail, became the inspiration of an entire string of "grand unified theories." Current GUTs and super-GUTs draw on quantum as well as on relativity theory and include four universal fields of interaction: the strong and the weak nuclear forces, in addition to electromagnetism and gravitation. The physical universe, it is assumed, obeys the laws of relativity as well as of quantum mechanics.

Grand unification takes elementary particles as elements within the four universal fields. Field intensity at a specific point gives the statistical probability of finding a particle at that location: in a sense, particles are generated by variations in field intensity. Photons, electrons, nucleons, and the entire particle zoo are consequences of the quantum dynamics of these interacting, physically real fields.

The above concept has produced a profound shift in emphasis in physics, from particulate entities to the ensemble of the dynamical matrices—fields—in which they are embedded. Nobel laureate U.S. particle physicist Steven Weinberg did not hesitate to assert that the real furnishings of the universe are fields; particles must be reduced to the status of epiphenomena.

Classical fields were soon complemented by quantum probability fields, first created in the 1920s and '30s by the European physicists Jordan, Wigner, Dirac, Born, Pauli, Fermi, and Heisenberg, among others. Quantum electrodynamics (QED) appeared during the 1940s. Its predictions found spectacular confirmation in the high-energy experiments that came on line in mid-century. As physicists succeeded in explaining disparate processes with field concepts, other quantum field theories followed, marking various stages in the unification of the physical forces of nature.

The first breakthrough came with the unification of the weak nuclear force with electromagnetism. Up to that point, the weak force appeared to behave in a very different way from that of electromagnetism. Sidney Sheldon, Steven Weinberg and Abdus Salam could show that these two forces constitute different manifestations of a single "electroweak" force. It is now believed that in the early moments of the universe there was no distinction between electromagnetism and the weak nuclear force. But as structure began to emerge in the universe, this perfect symmetry became broken and the integrated force differentiated into the long-range electromagnetic and the short-range weak nuclear force.

Further unification could be effected with a deeper understanding of the strong nuclear force. Before the advent of quarks it was assumed that the exchange of intermediate force particles (mesons) produced the effect of the strong nuclear force. However, with the quark theory of hadrons it was necessary to postulate a force between the quarks themselves. It turned out that this force could be mathematically treated in a totally analogous way to the force of electromagnetism. Although the force acting between quarks had yet to be unified with the electroweak force, its formal appearance was very similar. By analogy with quantum electrodynamics (QED), the theory that achieved this unification came to be called quantum chromodynamics (QCD). Thanks to these theories the number of the fundamental fields and forces in the universe could be reduced to just two: the integrated electroweak-cum-strong nuclear force, and the force of gravitation.

The first stage in the program of grand unification—the development of an integrated theory of the strong nuclear and the electroweak forces, together with the leptons and hadrons that constitute matter in the universe—had then to be completed with the next stage. This was to extend this grand unified theory to include the force of gravitation. This was to go beyond grand unification, to "supergrand unification."

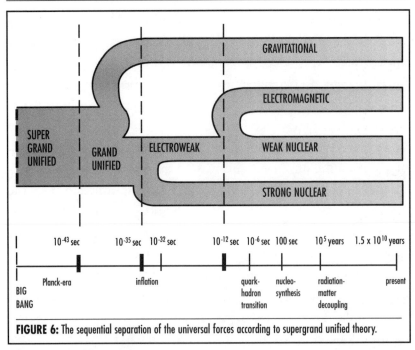

FIGURE 6: The sequential separation of the universal forces according to supergrand unified theory.

Supergrand unification requires the quantizing of the gravitational field. The strong nuclear force had been quantized in terms of the gluon; and the electroweak force had been expressed in terms of the W and Z particles. Physicists hypothesized that the gravitational force could be quantized in reference to a particle called the graviton.

Quantizing the gravitational field raised complex problems. Since Einstein's theory of gravitation is a geometrical theory of space-time, quantizing that theory means quantizing a geometry. Beyond this conceptual hurdle there are other difficulties. For one thing, there is no evidence that gravitons would exist in nature; for another, the mathematics that is needed to describe gravitons leads to infinities. As a result a quantum theory of gravitation requires a new approach, departing radically from earlier formulations of quantum field theory. So-called gauge symmetries have to be invoked, making use of supersymmetries and superspaces. In consequence a new generation of "super" theories was born.

Goal of Life is ... Ascension
"   "   "   " ... Levitation (personal)

The first breakthrough was the development of the mathematics of supersymmetry. The quantum field theory incorporating supersymmetry (known affectionately as Susy) became "quantum supergravity": it enabled physicists to unify fermions and bosons. This was a major achievement since the half-integral spin fermions are the main matter-particles, while the integral spin bosons are the particles of the universal forces. (Fermions and bosons are distinguished by the values of their "spin": bosons have integral values of spin [1, 2, 3 . . .] while fermions have half-integral values [$1/2$, $3/2$ etc.].)

Previously, the fermions could be grouped together in families, and so could the bosons, but there was a clear-cut separation between the internally related family of fermions and the likewise internally related family of bosons. Now, thanks to Susy, fermions and bosons—matter and force—could be related one to the other. In the higher dimensions of superspace, each could be "reflected" into the other.

In order to unify fermions and bosons in superspace, a whole new set of particles had to be introduced: for every fermion and boson there had to be a supersymmetric partner. Just as photons acquired mirror-image particles called photinos and quarks acquired squarks, the graviton had to be coupled with the supersymmetric gravitino. This resolved the principal obstacle to the unification of gravitation with the grand unified force. Theoreticians could now postulate a supergrand unified force: supergravity.

But supergrand unified theories encountered difficulties. First of all, quantum supergravity demanded that the supersymmetric partners should have masses higher than the particles of which they are the mirror images. This introduced an element of unverifiability into the theory: the energy level of the supersymmetric particles turned out to be high enough to preclude creating them in particle accelerators. The new particles would remain unobservable—unless, as some physicists believe, photinos can be detected in high-energy collisions between electrons and positrons, or protons and antiprotons.

Not only did the super-GUTs predict a host of new and experimentally unobservable particles, they also held another surprise: in most formulations they required 11 dimensions to work. Einstein's revolutionary innovation of adding the fourth time-dimension to the three spatial dimensions paled in comparison with theories that added as many as seven dimensions to the four of space-time.

Physicists went to work, using complex mathematics to "compactify" the seven extra dimensions of superspace in order to render Susy consistent with the four dimensions of relativity theory. It was assumed that the extra dimensions would exist but were "rolled up" so that their effect would not be manifest even at the scale of elementary particles. But it soon appeared that the effort was doomed: there was no way to reduce seven of the eleven dimensions without compactifying the remaining four as well. This, however, would reduce the empirical entailments of the theory to zero dimensions—a vexing development for a theory that claims to describe an aspect of physical reality.

For a time it appeared that the enterprise of supergrand unification had to be abandoned. But then a younger generation of physicists came up with another, even more daring idea. Joel Scherk proposed that particles are not particulate at all but strings that spin and vibrate in space. All known phenomena of physical nature would be built from different combinations of these vibrations, much as the music of a string quartet is built from the vibrations of the strings of the musical instruments.

The idea that rotating and vibrating strings would be fundamental to our understanding of nature dates back to the 1960s. At that time Gabriel Veneziano suggested that when elementary particles are arranged in order of their masses, they form a pattern similar to that of notes or resonances. Other physicists were later struck by the idea that the resonances could be produced by minuscule vibrating strings the size of particles.

Scherk's string theory proved compatible with Gell-Mann's quark theory. The new theory explained why quarks are unobservable in nature: it is for the same reason that a string can never have a single end. When the ends of a string are separated, new ends are created. In the same way, when hadrons are broken open, instead of single quarks, freshly paired quarks come into being.

In 1976, Scherk, Ferdinando Gliozzi and David Olive showed that supergravity can be introduced into string theory, making it into superstring theory. Here the particle-strings vibrate in a higher-dimensional superspace. But the real triumph of the theory came in the mid-1980s, just when supersymmetry theories seemed defeated by the problem of compactification. John Schwartz and Michael Green were able to show that a ten-dimensional superstring theory was perfectly compatible with four-dimensional space-time; it did not encounter the problems of compactification. The new super-strings turned out to be smaller than the strings of the original theory: they are no larger than the Planck-length of 10^{-35} meters— far smaller than any known elementary particle.

IN SUMMARY . . .

Unification in physics is a time-honored and bone fide endeavor. Its current form, grand unification, chalked up notable successes. This has opened the door to a still more "grand" form of unification: supergrand unified theory. Though super-GUT is in an early phase of development—certain aspects of it, such as superstring theory, are still beset by serious problems—confidence that grand, and ultimately supergrand, unification is achievable has grown. Few particle and field physicists would still contest that, in time, all the forces and particles of nature will be unified within a single theory.

This confidence is all the more remarkable as it could never be supported by actual experiment. To observe the electroweak force, born of the union of the weak nuclear force with electromagnetism, one has to generate 90 GeV (where one GeV [giga-electron volt] is the energy needed to create one proton). This level of

energy is just within the compass of current 100 GeV particle accelerators. Thus when such accelerators came on line and the unified
force appeared in the smashing of highly accelerated particles,
Weinberg and Salam were duly awarded the Nobel Prize. But the
unification of the strong nuclear force with the electroweak force
calls for about 10^{14} GeV, and an accelerator that could develop that
kind of power would have to be the size of the solar system. And
the supergrand unified force that would wed the grand unified
force with the force of gravitation requires 10^{19} GeV, the ultimate
scale of energy present in nature at the moment of the Big Bang.
To re-create that level of energy a particle accelerator would have
to be 100 trillion times more powerful than the Superconducting
Supercollider, the largest particle accelerator ever projected (but
never funded). Such a device would require a circuit 100,000 light
years long: an accelerator as big as our galaxy.

The fact is that the new physics has grown entirely beyond the
range of observable phenomena. No physicist ever hopes to find a
superstring "in nature," any more than a quark. A black hole cannot
be "seen," even if one were facing it, and an electron is bound to
remain a blur at the highest possible magnification: it describes a
million revolutions around the nucleus by the time a single light
quantum is emitted. In the same way, the unification of the forces
of nature remains a theoretical postulate, underwritten only by the
coherence, precision and elegance of the equations that specify it.
Yet this fact does not deter scientists from viewing GUTs and super-
GUTs as likely descriptions of the real world—a world that is believed to be deeper and wider than the world of our bodily senses.

Notwithstanding the growing level of abstraction of the theories, the unification of the physical world proceeds apace. Its current stand, though subject to revision and capable of improvement,
is already a stupendous achievement, a veritable milestone in
humankind's perennial search for a unified theory of the basic
contours of the world around us.

CHAPTER 10

THE SEARCH FOR UNIFIED THEORIES: 2. ACROSS THE DISCIPLINES

CIENCE'S SUCCESS in resolving the puzzles and clarifying the mysteries that beset its current world picture does not stand or fall with grand unification in physics. As one might expect, grand unified theories within physics apply only (or mainly) to the domain of physical nature. That is a limitation, however: physics covers a large chunk of natural phenomena, but by no means all. Evidently, matter (or matter-energy) not only coheres within particles, atoms and molecules, but also structures itself within cells, organisms and ecosystems—at least on our planet. Yet, even if, as Stephen Hawking noted, the goal of physics is a complete understanding of everything around us, including our own existence, it has not succeeded in reducing chemistry and biology to the status of solved problems, while the possibility of creating a set of equations through which it could account for human behavior remains entirely remote. While the physicists' grand unifed theories describe the properties and interactions of particles, atoms and molecules, they do not show how the existing particles, atoms and molecules generate the phenomena of the biological—not to mention the human—world.

A truly unified science would encompass *all* facets of the natural world, however, the physical as well as the biological, and even the neuropsychological. It would account for the progressive build-up of ever more complex and integrated systems with ever more differentiated characteristics, regardless of whether the systems belong to the field of physics, or to those of biology and the human sciences.

Can this challenge be met?

Discovering the interactions that could build the known universe toward the pinnacles of complexity where life appears, and then mind, is a staggeringly ambitious project. Nevertheless, there are pioneering scientists who contemplate tackling it. Forming an idea of their work can provide us with a valuable glimpse of the kind of transdisciplinary theorizing that could create a truly grand unified theory of the physical, as well as of the living, world.

BOHM'S IMPLICATE ORDER

English-born physicist David Bohm is perhaps the foremost pioneer of a transdisciplinary theory which is rooted in physics but not limited to the physical universe. His ideas have gained wider popularity than those of any other modern-day scientist, with the exception of his erstwhile mentor, Albert Einstein. They are discussed in scientific circles as well as among young people, even in alternative cultures and New Age circles.

Bohm's concept is of a basic simplicity and beauty, although it is quite radical. There are two levels or dimensions of reality: one that reveals itself at the surface, in physical and biological phenomena, and another at a deeper level that we can know only indirectly. A sound description of the universe must include the underlying level; Bohm named it "implicate" (meaning folded inward).

The essential feature of the implicate order is that everything that takes place in space and time—in the "explicate" order—is enfolded in it. An example is a vortex. It has a relatively constant,

recurrent and stable form, yet it does not have an existence independent of the movement of the fluid in which it appears. The
vortex may appear as an independent body, yet its order is derived
from the dynamics of the flowing water. In the same way particles
appear as independent entities, yet they derive from the underlying
"enfolded" order.

Bohm illustrated this principle with a device (actually built at
the Royal Institute in London) consisting of two concentric glass
cylinders with a viscous fluid such as glycerine between them. A
droplet of insoluble ink is placed in the fluid and the outer cylinder
is slowly rotated. The droplet becomes drawn out into a thread-like
form. If a sufficiently large number of rotations are made, the droplet is diffused and seems to be lost within the glycerine. If two
droplets are inserted in the fluid, each constitutes an independent
thread-like form; and if the threads intersect, the particles in each
droplet intermingle. Yet when the motion of the underlying fluid is
reversed, the particles in each of the threads retract back into separate droplets. Bohm points out that the carbon particles of the ink
are part of the total system—the glycerine solution—and are
enfolded or "implicated" in it.

The more technical aspects of Bohm's theory are not widely
known, yet they are significant: they concern the interaction of the
implicate and the explicate order. In the observable world of the
explicate order, the motion of matter-particles is constantly guided
by the implicate order. This guidance occurs through a "pilot wave"
called the quantum potential (denominated by the symbol Q).
Much as the gravitational constant G, the quantum potential pervades space-time. Q, however, originates in the implicate order,
which is beyond space and time. Thus the particles themselves do
not possess both corpuscular and wave properties: they are truly
particulate. The observed wave properties follow from the effect of
the implicate order-based pilot wave on their particulate structure.

Through the action of the quantum potential, the totality of the manifest world derives from the implicate order as an explicate subtotality of stable recurrent forms. Because all things are given together in the implicate order, there are no longer any chance events in nature; everything that happens in the explicate order is the expression of order in the implicate realm. Quarks as well as galaxies, the same as organisms and atoms, are once and for all part of the order that subtends the world of observation and experience.

HEISENBERG'S QUANTUM UNIVERSE

We now look at another theory that attempts to integrate what we know of the physical universe with our more immediate experience of life and mind. This is the legacy of Werner Heisenberg, revived and expanded by U.S. quantum physicist Henry Stapp.

Heisenberg himself was ambiguous about the philosophical consequences of quantum theory: at times he implied a mentalistic, at other times a physicalist, interpretation. He would write, for example, that "we are finally led to believe that the laws of nature that we formulate mathematically in quantum theory deal no longer with the particles themselves but with our knowledge of the elementary particles. . . . The conception of the objective reality of the particles has thus evaporated . . . into the transparent clarity of a mathematics that represents no longer the behavior of the elementary particles but rather our knowledge of this behavior." Yet Heisenberg would also maintain that "if we want to describe what happens in an atomic event we have to realize that the word 'happens' . . . applies to the physical not the psychical act of ob- servation, and we may say that the transition from the 'possible' to the 'actual' takes place as soon as the interaction between the object and the measuring device, and thereby with the rest of the world, has come into play; it is not connected with the act of registration of the result in the mind of the observer."

Evidently, if the transition from the "possible" to the "actual" (that is, the "collapse of the wave-function") is due to the interaction between the measuring device with a particle, the quantum world to which our observations refer is physically real. If, however, the wave-function collapses with the registration of the result in the mind of the observer, the quantum world beyond our observations is essentially mental. The former alternative gives us the so-called "ontological" interpretation of quantum mechanics, in contrast with the "mentalistic" (or idealistic) stance of the latter—the hallmark of the Copenhagen school.

Stapp chose the ontological interpretation (though he gives it an idealistic twist), and expands its application beyond the quantum domain to the domain of macroscopic phenomena. This yields the "Heisenberg quantum universe," complete with large-scale nonclassical effects.

The quantum universe dispenses with Bohm's implicate-order-based quantum potential, while retaining the idea that the probability distribution that occurs in quantum theory exists in nature, and not just in the mind of the observer. The quantum probability distribution, together with its abrupt changes, makes for a complete representation of reality. This representation discloses that the evolution of the physical world proceeds by an alternation between two phases: a gradual evolution via deterministic laws that are analogous to the laws of classical physics; and the periodic occurrence of sudden, uncontrolled quantum jumps. The latter actualizes one or another of the various macroscopic possibilities generated by the deterministic laws. The "detection event" (the interaction that collapses the wave-function) occurs in the context of a situation where deterministic laws have decomposed the quantum probability distribution into well-separated branches. This actualizes one of the alternatives and eliminates the others. In the Heisenberg quantum universe the actualized alternative is not limited to the microscopic world; it may also be a macroscopic event, distinguishable at the level of direct observation.

According to Stapp, the Heisenberg quantum universe gives a

coherent quantum mechanical explanation of biological, and even of mental, phenomena. In this universe the evolving quantum state, although it is controlled in part by mathematical laws that are analogous to the laws of classical physics, does not refer to anything substantive; it only represents the potentialities and probabilities associated with actual events. In consequence the universe is no longer matter-like: it is mind-like. The matter-like aspects of phenomena are limited to certain (nonclassical) mathematical properties, and these properties can be understood just as well as characteristics of an evolving mind-like world. Reversing the implications of classical physics—where there is no natural place for mind—is the Heisenberg quantum universe, where there is hardly a natural place for *matter*.

Stapp concluded that if these nonclassical mathematical regularities are accepted as characteristics of an essentially mind-like world, we "appear to have found in quantum theory the foundation for a science that may be able to deal successfully in a mathematically and logically coherent way with the full range of scientific thought, from atomic physics, to biology, to cosmology, including also the area that had been so mysterious within the framework of classical physics, namely the connection between processes in human brains and the stream of human conscious experience."

PRIGOGINE'S DISSIPATIVE SYSTEMS

Stapp's idealistic concept of the Heisenberg quantum universe requires us to relinquish our belief in a "matter-like" universe. There are other theories, however, that do not exact such a sacrifice. Alternative forms of transdisciplinary unification are possible, forms in which the physical world remains essentially matter-like. In these theories both mind and life are emergent factors resulting from the ongoing evolution of the universe.

Theories that take evolution as the key to transdisciplinary unification proceed on the observation that nature builds complexity in the course of time. Its developmental processes are sequential and ongoing, though they may be sudden and nonlinear.

Elementary particles build into atoms and atoms build into molecules and crystals. Molecules in turn build into macromolecules and into still more complex cellular structures associated with life; and ultimately cells build into multicellular organisms, and these again into social and ecological systems. It is not necessary, nor indeed reasonable, that each of these processes of assembly should obey categorically distinct laws. The same basic laws, functioning as nature's algorithms, could create the interactive dynamics whereby complexity builds in the universe from the level of particles to that of systems of organisms. These would be the basic laws of evolution in all of the natural realms. The transdisciplinary unified theories stating them would be general theories of evolution.

Until the last few decades, general theories of evolution were produced by philosophers who complemented the lacunae of scientific knowledge with speculative insight. Yet, although speculative, works such as Bergson's *Creative Evolution,* Herbert Spencer's *First Principles,* Samuel Alexander's *Space, Time and Deity,* Teilhard de Chardin's *The Phenomenon of Man,* and Alfred North Whitehead's *Process and Reality* stand as enduring milestones of evolutionary thinking. More recently concepts and theories have been developed that lift evolution as a general phenomenon from the realm of philosophical speculation into that of scientific investigation. This variety of unified theory was pioneered by Ilya Prigogine in his work on the thermodynamics of nonequilibrium systems and irreversible processes.

Prigogine was among the first to realize the unifying implication of the study of evolutionary processes. A living system, he said, is not like a clockwork that can be explained by simple causal relations among its parts; in an organism each organ and each process is a function of the whole. A similar point of view, he added, is necessary in the social sciences. The theory of the irreversible evolution of thermodynamically open systems applies to physical chemistry, to biological systems, and even to human systems.

To understand the thrust of this theory we should recall that classical thermodynamics was concerned with the transformation of free energy into waste heat in closed systems, with the consequent breakdown of order into randomness. In 19th-century physics, the ultimate implication of this line of thought was the running down (the "heat-death") of the entire universe. But since the first half of the 20th century, scientists have been exploring new approaches. Lars Onsager's 1931 study *Reciprocal Relations in Irreversible Processes* was pointing in the direction of irreversible processes that move systems away from, rather than toward, thermodynamic equilibrium. Then, in 1947, Prigogine devoted his doctoral dissertation to the behavior of systems far from equilibrium, and in the early 1960s Aharon Katchalsky and P.F. Curran elaborated the mathematical basis of the new science of non-equilibrium thermodynamics. These investigators showed that by concentrating on gradual changes in closed systems, classical thermodynamics had failed to confront real-world systems. The real world is populated by nonequilibrium systems that evolve nonlinearly and are open to flows of free energy in their environment. They import negative entropy from their environment (the physical measure of the order within a system that signifies the presence of free energy), and they export, that is, "dissipate"—entropy (used up energy). Such systems are basic to life: as Schrödinger noted in mid-century, "life feeds on negentropy."

Since open systems far from thermodynamic equilibrium dissipate entropy as they perform work, Prigogine named them dissipative systems (or dissipative structures). Such systems can be in a stationary state (when the negentropy they import from the environment precisely offsets the entropy produced within them), but they can also grow and complexify (if their negentropy import exceeds the entropy produced by irreversible processes within the systems). Dissipative systems only run down if their free energy import fails to match their internal entropy production.[1]

The dynamics of dissipative systems provides a basis for understanding the progressive complexification of nature. The complexification process gets under way when a critical fluctuation, either within the system or in its environment, destabilizes a system in a state far from equilibrium. The destabilized system either re-establishes a new dynamic equilibrium between its negentropy inflows and its own entropy production, or it enters a state of chaos that leads to its transformation—if not to its dissolution. If the dissipative system succeeds in finding a new state of dynamic equilibrium, we have a statistical probability that this state will involve more structure and complexity than the destabilized state it leaves behind. This means that in a population of dissipative systems instabilities, induced by random fluctuations, push a significant proportion of systems ever farther from the inert condition of thermodynamic equilibrium, toward the remarkable yet inherently unstable dynamic equilibrium where life—and perhaps mind—appears.

Perturbations, the random interplay of critical fluctuations, and the transformation that follows upon the destabilization of the previous system state, are the key elements that define the interactive dynamics of Prigogine's transdisciplinary unified theory. This dynamics governs systems in all realms of observation: the physical, the chemical, the biological, the ecological, and even the human.

IN SUMMARY . . .

Bold attempts at building unified theories—theories that unify findings within a given field of scientific inquiry—are successfully pursued in the new physics. They are joined by even bolder attempts to create transdisciplinary theories that hope to unify findings across a wide spectrum not only of the physical, but also of the biological and the human, sciences.

In regard to transdisciplinary unification, theories fall into two general categories: those that attempt to enlarge the laws and the scope of physics to embrace the living world; and those that investigate the dynamics whereby the physical world evolves sequentially toward phenomena that are no longer purely physical. Unified theories of the latter, evolutionary variety promise to convey a new understanding of how the physical universe produced life, and then mind, as part of its ongoing process of self-organization and self-creation.

NOTES

1. In terms of the basic formalisms of Prigogine's dissipative systems theory, change in entropy is defined by the equation $dS = d_i S + d_e S$, where dS is the total change of entropy in the system, $d_i S$ the entropy change produced by irreversible processes within it, and $d_e S$ the entropy transported across the system's boundaries. In an isolated system dS is always positive, for it is uniquely determined by $d_i S$, which necessarily grows as the system performs work. However, in an open dissipative system $d_e S$ can offset the entropy produced within it, and may even exceed it. Thus in a dissipative system dS need not be positive: it can also be zero or negative.

CHAPTER 11

UNIFIED THEORIES: THE STATE OF THE ART

T HE BREADTH and depth of contemporary unified theorizing appear impressive. But how accomplished are the theories in actual fact? We should attempt an assessment, first of the physics-based GUTs, and then of the transdisciplinary variety of unified theories.

ACHIEVEMENTS AND SHORTCOMINGS OF GUTs

In regard to GUTs, we should note that physicists have produced results that are truly remarkable, both for their scope and for the mathematical precision with which they are stated. A new picture of the universe is emerging; a highly unified picture. In this picture the particles and forces of the universe originate from a single "supergrand unified force" and, although separating into distinct dynamical events, continue to interact. Space-time is a dynamic continuum in which particles and forces are integral elements. Every particle, every force, affects every other. There are no separate forces and things in nature, only sets of interacting events with differentiated characteristics.

A focus on the basic or rock-bottom level of reality has proven to be unnecessary baggage left over from a classical theory that attempted to explain all things in reference to a combination of the properties of the ultimate building blocks, which it claimed were atoms. Today, a coherent and consistent set of abstract and mainly unvisualizable entities has replaced the notion of billiard-ball-like atoms moving under the influence of external forces. Processes in the physical world are no longer referred to the laws of motion governing the behavior of individual particles. Physical reality is not explained in terms of groups of fundamental entities, even if the entities are not atoms but quarks, exchange particles, super-strings, or yet to be discovered still more abstract units. This is important, because it is unlikely that phenomena of the level of complexity typical of life could be described by equations that center uniquely on the universe's smallest building blocks, no matter how precisely their motions could be computed.

The picture of an interacting and self-organizing universe is likely to remain valid, notwithstanding the high rate of attrition of the theories that expound it. It is difficult to see how physics would ever progress backward to a universe of separate material things and dynamical forces, to a mosaic of unrelated events in external equilibrium.

On the negative side, we should note that, though on the tech-nical score GUTs are remarkably accomplished, their scope and meaning have not been properly articulated. Scientists have been too intent on producing the mathematics that would unify the phe-nomena they observe to have ventured deeper into the implication of their formulas, while philosophers, the traditional interpreters of the knowledge of their times, have mostly kept away—with few exceptions, they have not caught up with the latest developments.

The lack of deeper thinking is showing. In the first flush of success some scientists have claimed that their grand unified theories can explain well-nigh everything. But in regard to grand and supergrand unified theories in physics, the label "theory of everything" (TOE) is clearly an exaggeration.

As we have seen, GUTs cannot satisfactorily explain the pro-gressive structuring of matter in space and time. Yet, as we have also seen, a theory capable of stating the laws that govern the progressive build-up of structure and complexity in the universe is possible, at least in principle. The question is whether such a theory can be formulated by enlarging the laws of physics, or whether it calls for transcending them in some way. Evidently, the more complex domains of nature are no longer domains of *physical* nature; theories in physics, as traditionally conceived, do not embrace them. But current theories in physics could perhaps be generalized—or, if necessary, completed with additional factors—so as to embrace the transphysical domains as well.

ACHIEVEMENTS AND SHORTCOMINGS OF TRANSDISCIPLINARY UNIFIED THEORIES

Among unified theories that embrace both the physical and the transphysical domains, David Bohm's ideas have pride of place. Bohm, we have seen, sought to complete the concepts of quantum physics with a further component: the factor Q, a pilot wave that originates in the most fundamental dimension of the universe, the implicate order. It is from here that the explicate order is generated.

The implicate order contains in a holographically enfolded form all that exists in the explicate order; the movement of enfolding-unfolding into and out of this order Bohm calls the holomovement. The entities we meet with in our experience are elements of the implicate order; they are brought forth by the holomovement. Even if such entities appear stable, independent and autonomous, they are as much a feature of the implicate order as a vortex is a feature of the flowing movement of a fluid. "The explicate order . . . is only an approximation and it cannot be properly understood apart from its ground in the primary reality of the implicate order, i.e. the holomovement. All things found in the explicate order emerge from the holomovement and ultimately fall back into it."[1]

It is but one step from the notion of a fundamental vacuum activity to the implicate order, and another step to the explicate order. Vacuum activity generates a flux in the implicate order, and the invariant features of the flux constitute the explicate order. That what remains in the background structure—the noninvariant features of the flux—is the implicate order. The problem with this view is that, even though the holomovement and the implicate order are the primary reality, we cannot have empirical experience of them, all our observations refer to the explicate order—the secondary reality.

The two orders as distinct realms of reality recall the classical philosophical problem of "primary qualities" (such as extension in space and duration in time), and "secondary qualities" (such as color and sound). Primary qualities, like Bohm's implicate order, are the true reality, yet we do not and cannot know them; secondary qualities, similarly to Bohm's explicate order, are the known world of observation and perception, yet they are not fully real. The latter, in Alfred North Whitehead's picturesque phrase, is the illusion and the former the dream.

The Heisenberg quantum universe championed and interpreted by Henry Stapp has other problems. When we generalize the laws of quantum physics to macroscopic phenomena we get a world where determinism and chance alternate. Deterministic laws create real alternatives, and the interaction of a particle with any part of the rest of the world selects one alternative and eradicates the others. This suggests a universal process that promises to account for the behavior not only of particles but of complex systems such as living organisms and conscious brains. But this promise is not entirely fulfilled.

One problem is that there are no laws in the theory that would explain how the quantum event "chooses" from among the available alternatives. The selection process remains unexplained; it is entrusted to chance. This introduces an element of unmitigated randomness that does not mesh with observations in the empirical world.

There is also another problem: in the macrodomain to which the theory is said to apply, there is an incessant process of

"choosing"—as a result, the wave-function of every object should
be constantly collapsing. While we can perhaps agree that a photon traveling from the light source that emits it to the counter that
registers it is free of interaction (and hence is in a probabilistic
quantum state), it is by no means clear how an organism or another
macroscale system could be sufficiently isolated from its environment to be in a similarly "pure" state. When such a system is in an
impure state—that is, when it interacts with any part of its environment—its wave-function must collapse. In the Heisenberg quantum
universe this would occur practically all the time. In matter-dense
regions such as our world, "decision-events" would be dense
enough to prevent dynamical laws from creating the alternatives
among which they are called upon to decide.

Could it be that the venture of expanding the theories of physics to embrace the world of life and mind is intrinsically fraught
with problems and paradoxes? If so, perhaps the evolutionary
variety of unified theory offers a better alternative.

Prigogine has disclosed the irreversible evolutionary dynamics
of systems in regimes far from thermodynamic equilibrium. Systems in this "third state" (far from, rather than in or near, equilibrium) behave in a remarkable manner: when destabilized by fluctuations they do not reach equilibrium but may restructure their
internal forces so as to take in, process and store more of the free
energies present in their environment. In consequence they do not
run down, but may wind up to reach increasingly dynamic regimes
and complex states.

Yet on closer scrutiny the thermodynamic enterprise, brilliant as
it is, also proves to be significantly (but perhaps not irrevocably)
flawed. Its problem is that when the evolutionary trajectory of a far-
from-equilibrium system bifurcates, what happens to that system is
at the mercy of chance. Prigogine cannot account for the system's
"choice" of a new dynamic regime following bifurcation anymore
than Stapp can account for the particle's "choice" of a deterministic
state when it interacts with the rest of the universe. In Prigogine's

nonequilibrium universe, the same as in the Heisenberg quantum universe, evolution remains punctuated by pure chance.[2]

This is a problem. If neither the past of the system nor the present state of the universe determines the outcome of a transformation, then the selection of the new dynamic regime in a system must be random. The way evolution unfolds in *one* system becomes unpredictable; and the way it unfolds in *many* systems is likely to be diverse. If in their evolution dissipative systems were driven by a random dynamics, they would tend to diverge and diversify. This is because any two systems, even if they were to start in the same state and with the same initial conditions, are bound to be exposed to different external influences and different patterns of internal fluctuation. Thus they are bound to embark on different pathways of evolution. It is not without reason that Prigogine spoke of a "divergence property" that would be basic to the evolutionary process.[3]

But if evolving systems were mainly to diversify over time, we would be surrounded by an uncoordinated welter of highly differentiated systems, instead of the remarkably consistent orders exhibited by the structures and constants of cosmology and the processes of physics and chemistry—to say nothing of the quasi-miraculously coordinated systems discovered in the biological and ecological sciences. If theory is to be adequate to the facts, it must, in addition to the dynamics of *divergence,* describe the dynamics of *convergence.*

IN SUMMARY . . . AND A FIRST CONCLUSION

What conclusions can we draw from this assessment of the state of the art in regard to unified theories?

It appears that current attempts at building unified theories of the observable and experimentally testable world convey remarkable insights, yet they have not produced a definitive theory. The situation is far from hopeless, however: current flaws could, in principle, be overcome. There is no intrinsic reason why a theory capable of accounting for all the manifest phenomena of the physical and the living world could not be constructed—why science could not develop a vision of all (or nearly all) the things that populate the universe. Lessons drawn from theories that attempt this feat could point the way.

The principal conclusion is that general evolutionary theories are on the right tack, but have to go further. They need to confront the problem of chance and accidental random divergence. To do so, they have to recognize the existence of fields and forces of interaction other than gravitation, electromagnetism, and the strong and the weak nuclear force. There could be other forces and fields in nature besides these four. There could be super*weak* forces, for example, that are not registered by the instruments presently used by physicists. Yet such forces could act on the indeterministic denizens of the quantum world, and on certain highly complex and super-sensitive systems of the macrodomain as well—on genes, and human brains and nervous systems, among others. And their action could make a crucial difference when it comes to explaining how a universe of atoms and molecules—and stellar bodies made of atoms and molecules—could build probabilistically but not randomly, toward the higher forms and domains of organization that are occasionally superimposed on the atoms and the molecules.

A fifth, superweak field operating in the womb of nature could subtly interlink particles, atoms, molecules, cells, organisms, and entire ecologies. Linkage by this force could make the many systems that arise and evolve in nature not only diverge among themselves, but also converge within successively higher systems and metasystems, accounting for both the *diversity* and for the *order* that we find in experience. It will lead us to the discovery of the unity and consistency of the grand sweep of nature's evolution, as it builds from microscopic quanta to macroscopic organisms and, ultimately, to mind and consciousness.

The addition of a subtle "fifth field" will give science the key to a consistent and comprehensive account of well-nigh everything that evolves in the universe. This is a fascinating prospect. We shall explore it in the fourth and last part of this study.

NOTES

1. Bohm's rationale for equating "holomovement" with "implicate order" and naming them the primary reality originates in a chain of theoretical considerations that led him to the concept of the implicate order in the first place. As his collaborators Basil Hiley and David Peat note (in *The Undivided Universe,* Routledge, London 1993) Bohm wanted to abandon the traditional notion of particles and fields-in-interaction in a continuous space-time, and replace it by the notion of "structure process." In the topo-chronological approach he chose, the presence of matter and of fields is indicated by a break in the space-time background structure. Thus the world of matter is essentially a break in the wholeness of the cosmos—a wholeness rooted in the vacuum state, viewed as a background matrix full of undifferentiated activity. Vacuum activity is the basis; the physical world as presented to observation is but the setting for the quasi-stable, semi-autonomous entities that arise as a result of this activity. Even space-time is but a derivate from holomovement as vacuum activity.

2. According to Prigogine, the evolutionary process depends on an element of randomness in the system. "Only when a system behaves in a sufficiently random way may the difference between past and future, and therefore irreversibility, enter into its description," wrote Prigogine with Isabelle Stengers in *Order out of Chaos.* "The 'historical' path along which the system evolves as the control parameter grows is characterized by a succession of stable regions, where deterministic laws dominate, and of unstable ones, near the bifurcation points, where the system can 'choose' between or among more than one possible future. Both the deterministic character of the kinetic equations whereby the set of possible states and their respective stability can be calculated, and the random fluctuations 'choosing' between or among the states around bifurcation points, are inextricably connected. This mixture of necessity and chance constitutes the history of the system."

3. A universe dotted with random bifurcations also creates intractable

problems of computation. As one bifurcation gives rise to (at least) two trajectories, each of which bifurcates into two further trajectories, the number of trajectories grows as the exponential of the number of bifurcation series. After a series of only 100 bifurcations, there would be approximately 10^{33} trajectories (that is, systems with different evolutionary pathways). A superfast computer programmed to examine the possible trajectories could achieve about 10^6 computations per second. This would require 10^{27} seconds to check the trajectories (and corresponding states) resulting from merely the first 100 series of bifurcations. But 10^{27} seconds makes 10^{20} years—considerably more than the current estimates of the age of the universe. On the other hand, if the bifurcations are not random we do not need to compute all possibilities, only those that are probable. This would considerably reduce the computational work load—and hence time.

PART FOUR

THE EMERGING VISION

---

*We are seeking for the simplest possible
scheme of thought that can bind together
the observed facts.*

Albert Einstein, *The World As I See It* (1934)

CHAPTER 12

THE KEY TO "QTV"

SIGHTING THE FAR SHORE

P ART ONE of this book offered a grand tour of science's established vision of the world: a vision of the near shore. This vision is important for understanding as well as for action, but to the lay person it is often hidden beneath complex formulas and abstract concepts.

Our tour of discovery started with a review of what science and scientists understand under cosmos, matter, life, and mind, and noted that this is quite different from what common sense—and even mainstream public opinion—usually believes that it is. Then, in Part Two, it became evident that the dominant vision of science is not final and definitive but has a fair number of fuzzy areas and even some black holes. In Part Three we looked at the kind of innovation—"revolution" is a better word put forward by leading-edge scientists in their effort to create an integrated, consistent and coherent view of the whole of the reality that underlies our obser-vations and experiments. We have progressed in this way from the established vision of science to the problems that blur that vision, and then to the current quest for an unblurred, more comprehen-sive and unified vision—to the new, less problem-ridden shore.

We now go further. We embark on a tour of anticipation, an exploratory safari. Our destination is the currently glimpsed far shore; a shore that promises a coherent and consistent vision of the principle kinds of things that exist and evolve in this universe, including matter, life, and mind. Such a "quasi-total" vision (we shall call it QTV) is no longer utopian: it is the tangible fruit of detailed research and innovative theory-building as the current scientific revolution grows toward a fuller maturity.

Let us take a better look at the concept of QTV. Why is that vision just "quasi"? Could not scientists hope to graduate to an entirely complete, truly total vision of the observed and observable world? Unfortunately not: that would be overstepping the scope of science. A more modest (yet already incredibly ambitious) stance is fully warranted. First, because a truly total vision of the known world would include spiritual and metaphysical element, and these—intuitions of the divine, of soul, and of other transcendent realities—are not accessible to scientific scrutiny, either now or in the foreseeable future. Second, because the items of experience that *are* accessible to science form an open set. New items could always be added to it, much as quarks, black holes and superconductors were added in the recent past. At any given time even the seemingly most complete scientific vision is only quasi-complete in regard to the vision that could emerge at a later time.

Evidently, a quasi-total vision in science is already a tall order. It would integrate our knowledge of physical nature with that of living nature, and both sets of knowledge with our more intimate knowledge of mind and consciousness. Such integration is now within the realm of real possibility. It calls for establishing coherence among the items of knowledge that we already possess. This is feasible in principle.

As we have seen, the overall knowledge of the world we presently get from science has limited degrees of coherence. Within given fields of investigation there *is* coherence: to establish it is the real function of theories. What botanists observe when they look at plants is coherent with the system of classification worked out by Linnaeus, and it is coherent also with the current principles of biochemistry and plant biology. But it is less coherent with what scientists know about human physiology, and hardly coherent with what they know about the internal structure of the atom. The greatest breaks in the coherence of the current scientific world picture arise between the physical world and the living world, and between these worlds ("nature") and the world of the conscious human mind. A genuine transdisciplinary unified theory establishes coherence also between these domains.

A genuine unified theory is an encompassing theory with quasi-total vision. It orders the existing elements of science's knowledge of the world and makes them rationally understandable. It enables us to know the world *better,* rather than to know *more* of it. Hence, rather surprisingly, instead of complicating scientific knowledge, a transdisciplinary unified theory simplifies it. This does not occur at the expense of detail and precision. For example, when psychologists know something about human nature in general, this does not prevent them from knowing more about the psychology of one particular patient. On the contrary; by relating the patient's unique personality traits to basic insights about the common features of human personality, they can come to a deeper understanding of his or her problems. After all, what is unique about a human being, or anything else in the real world, is not this or that trait (if it were, we would be entirely puzzled by it) but the *combination* of familiar traits that occur in regard to them. The same applies to organisms and quarks, and to all the denizens of the observable and knowable world.

A good scientific theory shows how unique features are specific combinations of in themselves nonunique elements. A good general theory takes this further: it establishes coherence among a larger number of (individually unique) things. And a genuine unified theory takes it further still: it creates coherence among *all* the things of which we have some variety of scientific knowledge.

How would science go about creating a unified theory with such QT vision? The most logical way to do this is to start with the remaining puzzles and paradoxes: the breaks in the coherence of the current scientific world picture. If scientists can come up with the key to overcome the puzzles and re-establish coherence the various items of existing scientific knowledge can be built into a consistent theory—into a cathedral of many and diverse elements, yet of harmonious overall design.

The question is: do the breaks in coherence have a common thrust, and thus a common resolution? If they do not, unification based on a general concept will be impossible. But if they do, scientists could find the key that would unlock their coherence. Let us have another look, then, at the persistent paradoxes of physics, biology, and the sciences of mind and consciousness.

The Persistent Paradoxes: A Short Catalogue

(A) *Paradoxes of the physical world*

Elementary particles in identical quantum states remain instantly interconnected even when separated by finite distances; photons emitted one by one interfere with one another as simultaneous waves; electrons in superconductors flow in a highly coherent manner taking on identical wave-functions; these particles are instantly and nondynamically correlated in different atoms even if they were not previously associated with one another, and they are specifically correlated in the energy shells that surround the atomic nuclei. Four different elements—helium, beryllium, carbon, and oxygen—are so precisely tuned in regard to their resonance frequencies that sufficient carbon can be produced in the universe to create the physical basis for life. And the universal constants themselves are so finely adjusted to one another that life can emerge on Earth, and conceivably on other planetary surfaces.

(B) *Paradoxes of the living world*

The morphology, and even the genetic information, of widely different species exhibit striking coincidences— even if evolution within its finite time-frame is believed to be governed by intrinsically random and disconnected processes of mutation and natural selection. Living species are able to generate and regenerate their highly complex form although each of their cells contains but one identical set of genetic instructions; and, if changes in the environment require basic changes in the adaptive plan of a species, those changes are produced on occasion through massive and highly coordinated—decidedly nonrandom—genetic mutations.

(C) *Paradoxes of the human mind*

Memory and inter- and transpersonal communication exceed the range that was traditionally assigned to the human brain and nervous system. In particular circumstances people appear able to recall any, and perhaps all, of their experiences, and possibly also the experiences of others; and on occasion they seem able to affect each other's mental and bodily states across space and time. And individuals as well as entire cultures seem capable of engaging in transpersonal contact and communication, sharing some of their ideas, artifacts and accomplishments beyond the range of the ordinary forms of personal and cultural interaction.

The burning questions that crop up in regard to these vexing (but also intriguing) paradoxes are at least the following:

- How could the universe at time zero anticipate the conditions that came about 10 billion or more years later?

- How can we account for the coincidence of the energy levels of four different nuclei?

- How can each photon pass through *both* holes in the double-slit experiment, even if it was emitted as a single particle of energy?

- How could one particle "know" the state of another particle — seeing that in superconductors, around shared nuclei, and even in discrete atoms, it assumes a corresponding state?

- How is it that, when exposed to major changes in their milieu, species that are finely tuned to the prior conditions of their environment manage to survive—instead of dying out and leaving the world populated mainly by algae and bacteria?

- How could nearly 40 phylogenetically distinct types of insects and animals have acquired one and the same master control gene for building an eye? Did they access the information from some archetypal form or pattern—or from each other?

- Why do organisms possess programs that can repair artificial damages inflicted by scientific curiosity in the laboratory, when these programs could not have been naturally selected in the history of their species?

- Where do "lifetime memories" and memories from apparent *previous* lifetimes come from? Can a 10-centimeter-diameter brain hold 2.8×10^{20} "bits" (or more) of information?

- How is it that as many as a quarter of all people—and not just sensitives—have the ability to "read" some aspect of the mind of the person with whom they interact?

- How can one person spontaneously and directly affect the body and mind of another person—perhaps even "see into" the other across considerable distances and tell what is wrong with him or her?

- Could it be that several people meditating together enjoy some kind of collective consciousness—and that the focused collective consciousness of a group of people affects the bodily condition of other people?

- And is it mere coincidence that different and widely separated cultures, as well as various branches of art and science, come up with striking parallelisms and "synchronicities" from time to time?

These questions do have a common answer, for the puzzles and paradoxes they refer to have a common thrust. *Everything we query here is possible, provided there are subtle and ongoing connections among the things and events that co-exist in the universe.* Given such connections, microparticles can be informed of each other's state within given systems of coordinates; the genome of living organisms can be linked with the relevant aspects of the environment; and human brains and minds can communicate with one another transpersonally, across space and time.

There is a space- and time-connecting factor in the various domains of nature: the physical, the biological, as well as the psychological. Because, in the absence of interconnectedness, nothing more interesting could come about in the physical universe than hydrogen and helium, the presence of complex systems such as those of life would have to be ascribed to an unfathomable stroke of luck, or the will of an omnipotent Creator. Likewise, the evolution of biological systems, and their generation and regeneration, would require explanation in terms of mysterious "building plans" or other metaphysical factors instead of bona fide scientific concepts rooted in observation and experiment. And if we do not recognize the possibility of spontaneous interconnection among human minds, many of the most fascinating aspects of human experience would have to be ignored, or dismissed as superstition and fantasy.

SUBTLE CONNECTIONS: THE BASIC CONCEPT

Subtle interconnections seem required for finding a meaningful solution to the wide variety of puzzles and paradoxes that beset the contemporary scientific world picture. This finding coincides with the preliminary conclusion we have reached in assessing the state of the art in transdisciplinary unified theories: that the processes of evolution become coherent when we assume that randomness is mitigated by a superweak force interconnecting the evolving systems. It appears, then, that the requirement for a significantly puzzle-free theory is the same as that for a coherent transdisciplinary unified theory. This is significant. The factor that clears up some of the most vexing problems in the contemporary natural sciences is the same as that which could unify the findings of these sciences. The current search for an interconnecting "fifth field" in nature seems entirely on target.

Before reviewing the latest findings of the fifth-field research, we should take a moment to clarify a question of principle. Just how would an interconnecting field cope with the vexing problems of chance in evolution?

We can best clarify this question by taking two intriguing examples as illustrations. They were advanced by world-class scientists, though they were originally intended to highlight different points.

The first example comes from astrophysicist Sir Fred Hoyle. Suppose, said Hoyle, that a blind man is trying to order the scrambled faces of a Rubik cube. As the experience of anyone who has tried it shows, matching the colors on all six faces of the cube can be a lengthy process; even a bright and physically nonhandicapped person can spend hours groping toward the solution. A blind man would take much longer, since he is handicapped by not knowing whether any twist he is giving the cube brings him closer to or further from his goal. In Hoyle's calculation, his chances of achieving a simultaneous color matching of the six faces of the cube are between 1 and 5×10^{18} twists. Consequently the blind man is not likely to live to see success: if he works at the rate of one move per second, he will need 5×10^{18} seconds to work through all possibilities. This length of time, however, is not only more than his life expectancy: it is more than any reasonable estimate of the age of the universe.

The situation changes radically if the blind man receives prompting during his efforts. If he receives a correct *yes* or *no* prompt at each move, he will unscramble the cube, on average, in

120 moves. Working at the rate of one move per second, he will need on average two minutes, rather than 126 billion years, to go through the moves that lead him to his goal.

Hoyle's calculations illustrate the difference that interconnections—in this case in the form of a constant feedback of information—make in a goal-seeking process. In this example the feedback prompting the player is perfect information: the prompt is always correct. If the information is less perfect (or less compelling for the player), there will be random mistakes and the player will take longer to reach his goal. Yet even an occasional and non-compelling "prompt" will speed up otherwise randomly groping goal-seeking processes.

In Hoyle's example the goal is given from the start: it is to match the colored faces of the Rubik cube. But in nature goals are not likely to be given ready-made. Scientists mistrust "teleology"—the assumption that nature follows a blueprint that was laid down when a process started. Instead, many scientists believe that the very process of seeking a goal generates it. How is that possible? Another intriguing example, this time from quantum physicist John Wheeler, suggests an answer.

Wheeler's example regards the popular parlor game known as "Twenty Questions." In this familiar game the object is to identify a particular object or person agreed upon by the players by means of a series of 20 questions to which only *yes* or *no* answers can be given. One person leaves the room while the others think up the object or person he or she is to guess. The guessing begins by asking general questions such as "Is it vegetable?" and then proceeds to more specific queries such as "Is it larger than an elephant?" In the final stages of a well-conducted interview a definite question can be posed, such as "Is what you have thought of the lamp on the street corner?"

In the usual variant, the game is goal-oriented: the players establish the thing or person to be guessed. But the game, says Wheeler, can also be played in another way. The players conspire not to think of any thing or person to be guessed, but not to disclose this to the one who does the guessing; he or she will ask questions as if there were something definite to find out. The game would end in utter confusion were it not for a simple rule that the players decide to obey: any answer they give must be consistent with the answers they gave before. If, for example, the answer to the question "Is it vegetable?" happened to be *yes,* all further answers must be given as if the thing to be guessed were a plant. As the questions move from the general to the particular, the range of permissible answers becomes progressively restricted. A skilled interlocutor can arrive at a specific question to which the other players, bound by the no-contradiction rule, will be obliged to answer *yes.* It turns out that the game reaches a specific goal, even though none was set at the beginning.

This particular example shows that "games" that remember their own past states and feed back the relevant information achieve a distinct goal-orientation. And they proceed toward their self-generated goal with far more speed and efficiency than a process based on random trials and errors.

In nature, these factors would make an almost magical difference. When interconnections feed back information on the past to processes in the present, the feedback limits the random play of probability in the evolution of complexity, speeding up developmental processes and rendering them self-consistent. The "divergence property" noted by Prigogine becomes complemented with a "convergence property"—all of nature turns into a goal-generating and self-evolving system. And the divergent/convergent orders achieved in the process come about within time-frames that do not exceed the time we have reason to believe was actually available for physical evolution in the cosmos, and biological evolution on Earth.

A theory that discloses the processes whereby the natural world feeds back information on the evolution of the things that exist in it could account for the way complexity unfolds from the Big Bang (or before) to our day. Ultimately, such a theory could account for well-nigh everything—given that everything in the universe is the result of an interactive process of self-creation. It would be a unified theory of the evolutionary variety, providing us with a quasi-total vision of the scientifically knowable universe.

CONNECTIONS IN SPACE AND TIME

Interconnections, we have seen, would work a kind of magic in nature: they would transform a randomly groping world into a self-consistently self-evolving universe, one that we could grasp through a single, highly general but self-consistent and hence potentially exact theory. But could such connections exist in the real world around us?

Let us explore now the real-world possibility of universal interconnections. Would they entail metaphysical or supranatural principles? We begin with the logic of connections in *space,* and continue with the possibility of connections in *time.*

In regard to spatial connections, we should note that if a thing or event at one point in space is connected with a thing or event at a different point, there must be something that transmits the effect from the first to the second. "Action at a distance" is not an acceptable notion: commonsense suggests that there is a continuous medium that stretches between, and hence interconnects, the two things or events. Scientists view continuous mediums of this kind as *fields.*

Fields are curious entities: ordinarily only their effects are observable, the fields themselves are not. In this regard fields are like a superfine net. If the threads of the net are thinner than the naked eye can distinguish, we do not see the net itself without suitable instrumentation; we can, however, see the knots where several threads come together. The knots appear to float in thin air, yet they are connected by the threads, so that when one knot moves, the others move as well. When we notice that the motion of one knot is connected with the motion of other knots, we have to assume that a correspondingly extensive net connects them.

Fields that interconnect phenomena can also be likened to a set of connected springs. As one spring is depressed, all the others become bent, depressed or expanded accordingly: the surface moves coherently, though not uniformly. This is a dynamic metaphor for the behavior of particles according to string theory. In that conception particles are localized vibration-patterns in continuous vibratory fields. The vibrations are connected through force fields, so that a change in the frequency of one vibration produces corresponding changes in the frequencies of the others.[1]

Nets and springs are good metaphors for so-called classical fields. These fields are *causal,* and *local.* Here causal means that the field produces completely predictable interactions: when a body is placed in such a field, it is always affected in exactly the same way. For example, when a bullet is shot in the gravitational field of the Earth, it always describes precisely the same parabolic curve. Local, in turn, means that changes in the field propagate at the speed of light, or below it. If the Sun, for example, were to suddenly vanish from its usual place in the solar system, the gravitational effects would not be felt on Earth for a full eight minutes— the time it takes for light from our star to travel to this planet.

However, there are also fields that are non-classical. These are known as quantum fields, and they are neither causal nor local. Objects such as particles in quantum fields do not have both definite positions and momentums: they are intrinsically indeterminate. They are also correlated in ways that transcend the limitations of signal-propagation at the speed of light. Quantum fields do not determine the actual state of the objects embedded in them: they merely state the *potentials* for the manifestations of physical effects. These potentials are inherently probabilistic. Quantum fields describe the behavior of physical objects that cannot be described in classical terms, that is, as obeying determinate causal laws and having single, determine locations in space.

Whether quantum fields are just theoretical devices, or whether they describe a true and uneliminatable indeterminacy at the heart of physical reality, is as yet an open question. Quantum physicists tend to lean toward the latter interpretation, whereas physicists who side with Einstein in this debate favor the former: they look on quantum fields as explanatory tools to be used for purposes of calculation until something better comes along. Here we shall leave the question open until we examine the possibility that a physically real "fifth field" may exist in nature. If it does, quantum fields could turn out to be effects or consequences of that underlying field (which, however, may prove to be non-classical in its own right: it could have properties that go beyond those of classical fields).

Let us go on, however. What about memory, that is connections in time? In classical physics, temporal connections between objects were thought to be conveyed not by fields, but by a continuous causal chain. Physicists traced observed effects to assumed causes by postulating universal laws of motion and rigorous chains of cause and effect. The initial conditions of every process were seen as the effects of prior causes that, in turn, were the effects of causes that were still prior. Thus an unbroken causal chain seemed to have stretched back to that hypothetical first instant when the universe was set in motion. The initial conditions that reigned at that instant were assumed to have determined everything that has taken place since then.

But this form of temporal connection is no longer affirmed by scientists. By the first decades of this century the determinism of classical mechanics had been discarded, and time-linkages through causal chains had been rejected. A probabilistic universe such as we have today cannot be "caused" by its past; at the most, specific events can leave traceable impressions on a limited range of subsequent events.

To understand how contemporary science conceives of the interconnection of things and events in time, we must apply a different analogy. If one event is linked with another in time, then we are dealing with memory: the former event is in some sense "remembered" by the latter.

On first sight, memory seems to restrict the discussion to the human mind. But on a second look, memory turns out to be a broad concept, with application to the physical and biological as well as the human world. This is because, while in humans memory is associated with mind, there are nonmental forms of memory in physical as well as in living nature. The simplest of living organisms conserves some impressions of its environment: it has some variety of memory although it does not possess a nervous system capable of mind and consciousness. Even an exposed film has memory: it "remembers" the pattern of light of various intensities that reaches its surface through the camera lens. And the computer that processes the text now being written has memory—and a form of logic and intelligence—though it is not likely to have mind and consciousness.

However, it is the type of memory associated with the hologram that is the most likely candidate for accounting for temporal interconnections of a universal kind in nature.

Consider the hologram. Basically, it is a wave-interference pattern produced by two intersecting beams of light stored on a photographic plate or film. One beam reaches the film directly, while the other is scattered off the object to be reproduced. The two beams interact, and the interference patterns encode the characteristics of the surface from which one of the beams was reflected. As the interference pattern is spread across the entire film, all parts of it receive information regarding the light-reflective surface of the object. This means that the hologram stores information in a distributed manner.

Because all parts of the hologram receive information from all parts of the photographed object, the full 3-D image can be retrieved by reconstructing the wave-interference patterns stored on any part of the film—though the smaller the part used in reconstructing the information the fuzzier the image. When two or more parts of the film are viewed simultaneously, observers on different locations retrieve the same information at the same time.

In addition to being distributed, holographic information storage is extremely dense: a small portion of a holographic plate can conserve an enormous variety of wave-interference patterns. According to some estimates, the entire contents of the U.S. Library of Congress could be stored on a multi-superposed holographic medium the size of a cube of sugar.

These properties of holographic information storage suggest that temporal connections in nature would very likely be in the form of a hologram. Nature would have a holographic memory.

Nature's holographic memory could not subsist in empty space—it would have to be based on a continuous medium that carries the hologram's interfering wave patterns. Hence we have to turn again to the concept of field: nature's memory must be based on a holographically information-conserving and transmitting field—a *holofield*.

How the universe's holofield could conserve and transmit information can be illustrated with the example of ships on the sea. Scientists have found that the surface of bodies of water—seas, lakes, or ponds—is highly information-rich. The patterns of waves disclose information on the passage of boats or ships, the direction of wind, the effect of shorelines, and various other factors that have previously perturbed the surface. The wave patterns may be conserved for hours, and sometimes for days, after the vessels themselves have disappeared. (We can see such patterns with the naked eye when the sea is calm and we view it from a sufficient distance, such as a high cliff or an airplane.) Indeed, native Polynesians have learned to navigate their craft by the patterns created by islands in their stretch of the sea. Though ultimately the patterns dissipate as the waves are eroded by the combined action of gravity, wind, and shorelines, as long as they persist they provide information on all the things that have taken place in a given region of the sea.

The wave patterns not only provide passive information on a stretch of the sea, they also actively if subtly influence everything that takes place on it. The effect of the waves created by one ship on the motion of other ships is ordinarily minor: in a large ship one hardly notices the pitch or roll induced by the wake of another. But it can also be dramatic, as anybody who has sailed a small boat behind an ocean liner can readily testify. This means that there is a subtle but effective transfer of information through interfering wave fronts; and this is very much the same whether it occurs on the surface of the sea or on a holographic film.

Even if directly unobservable, a continuous holofield in nature could ensure both spatial and temporal connections. Connections in space, we have seen, call for the simultaneous availability of information at different spatial locations, and the distributed nature of information in a holographic field can respond to this requirement. And connections in time call for the enduring conservation of a staggering amount of information, and a holofield can satisfy this requirement as well.

It is interesting (and important) to note that, given universal interconnections in space and time, there is no such thing as pure chance in the world. No thing or event is entirely disjoined from any other, hence even seemingly accidental "coincidences" have an underlying logic. This does not mean that all things would be bound together within the iron clutch of natural laws: the way things are connected may be extremely subtle: they may show up only with a statistical significance—the effects could emerge only when there are a large number of things involved, or a large number of tries. (In throws with a large number of biased die, even the most subtle weight differences at the six sides will produce a noticeable tendency toward the number at the opposite side—the same as in a large number of throws with one biased die.)

The assumption that there are subtle connections between any two things in the world means that all things are correlated in some way with all other things. This has a remarkable entailment for contemporary physics. As French physicist Costa de Beauregard observed, when two physically meaningful things occur together, we must assume a "covariant" propagation of information between them.[2] And if correlations extend throughout space and time, then information must be covariantly propagated throughout space-time.

A FIFTH FIELD

It is not enough to suppose that a space- and time-connecting holofield *could* exist in the real world; we must also ask, *does* it exist? If it does, there is an information-transmitting field extending throughout space-time. What would such a field be like? Following these concepted explorations, we turn to the findings of actual fifth-field research. What does it tell us about the nature of this field? Is it a classical field or a quantum field—or something different again? Let us look at the possibilities.

Hitherto science, we have already noted, recognized four kinds of universal fields in nature: the gravitational, the electromagnetic, and the strong and the weak nuclear fields. According to the GUTs of the new physics all four have originated as a single "supergrand unified force" in the very early universe. The currently observed fields have separated out by spontaneous symmetry-breaking in the rapidly expanding and cooling phase that followed the Big Bang. Could it be that we do not need a fifth field, since one of these fields has properties that make it into a universal holofield?

This is unlikely. The strong and weak nuclear fields are local forces of interaction; they could not interconnect phenomena across wide stretches of space and time. Gravitation and electromagnetism are cosmically extended fields, yet the kind of connections we have noted constitute anomalies in regard to the established theories. To accommodate information conservation and transmission, they would have to be transformed beyond recognition. As already hinted, it would make more sense to look further, to a superweak (but by no means ineffective or insignificant) "fifth field" that would operate in nature.

Though a fifth field is not—or not yet—part of the physicists' repertory of uncontroversially-known fields, a number of outstanding scientists have been hypothesizing that it might exist. Already in 1967, Harlow Shapley, the hard-nosed Harvard astrophysicist, asked whether there might be an "additional entity, a fifth one" in the universe besides space, time, matter, and energy. Would not this fifth entity be necessary if one had the assignment of creating the universe? How about Drive, Direction, Original Breath of Life, or Cosmic Evolution? Shapley settled on the last as the most likely. Cosmic Evolution, he noted, may be a fifth entity which we would need for understanding a dynamic universe.

We now see that to understand a dynamic universe we can make do with a more modest concept: a cosmically connecting holofield. This would be the superweak "fifth field," subtly interacting with the four known fields. Physicist William Tiller came to the very same conclusion. "In conventional science," he wrote, "four forces are considered responsible for all the observable phenomena in the universe: the strong nuclear force, the weak nuclear force, the electromagnetic force, and the gravitational force. However, a growing body of experimental data has appeared that seems inexplicable based on these forces alone." Tiller calls the force required to explain the otherwise inexplicable data "a field of subtle energy." As a subtle-energy field is not likely to be a classical field; on the other hand, it may not be a pure quantum field either. It is likely to have its own physical reality and its own physical properties, but these properties need not be the same as those of the known classical fields.

A number of physicists have speculated on the nature and existence of a field with subtle yet universal effects, among them, as we have noted, David Bohm. While the mainstream of the contemporary physics establishment has been reluctant to embrace the concept of such a field, more and more theories and hypotheses are now put forward to account for the paradoxical quantum correlations in terms of a physical real (as contrasted with a pure probability) field. The hypotheses, as we shall soon see, interpret space-time as a physically real field, or else look to the quantum vacuum that pervades space-time as the potential source of the field.

In biology the field concept had a longer if just as controversial history. To understand how the remarkably ordered forms of living nature could have come about, a number of biologists have suggested that, in addition to biochemical processes and genetic programs, a field of a specifically biological kind must be active in the organism.

The debate on biofields dates back to the 1920s, when Alexander Gurwitch postulated a morphogenetic (form-generating) field. He was led to the concept by noting that in embryogenesis the role of individual cells is determined not by their own properties or by their relations to neighboring cells, but by a factor that involves the entire self-organizing system. He postulated a system-wide "force field" generated by the particular force fields of individual cells. Though at first Gurwitch claimed that this overall field is nonmaterial, he later allowed that the concept could be translated into the language of physics.

The early biofield idea was elaborated by a number of biologists, including N.K. Kolciov in the former Soviet Union, Ervin Bauer in Hungary, and Paul Weiss in Austria. They noted many otherwise inexplicable phenomena, such as the spontaneous reassembly of the separated cells of a sponge, the regeneration of the limbs—and even of the iris of the eye—of the newt, as well as the ability of some species of fertilized ova to develop into the full organism even when the molecular substructure is destroyed. It was said that when a planarium (a type of flatworm) is cut in half, its regeneration into a complete organism is guided by its biological field. Just as when a magnet is cut in half two new magnets form, each with its own complete magnetic field, so when the worm is bisected its field splits into two identical biofields. Each of the fields guides the processes that regenerate the complete worm.

During the last 50 years field-like phenomena were discovered in various domains of biology, and such initial speculations under-went considerable development. D'Arcy Thompson produced a path-breaking work on the evolution of form in living species, il-lustrating it with continuous transformations in fish; Hermann Weyl demonstrated the self-consistent transformations of symmetry in the form of a vast number of organic species. Conrad Waddington and René Thom divided the biofield into geometrical zones of structural stability, connecting geometrical forms with dynamical processes in living systems. Yale biologist Harold Saxton Burr sug-gested that the biofield is an "L" (life) field that directs and orga-nizes the physical structure of the organism. Burr's collaborator Leonard Ravitz claimed to have uncovered evidence that the L-field disappears just before physical death.

More recently biologists such as Brian Goodwin argued that biofields are associated with growth processes in plants and ani-mals. According to Goodwin, the forms of living nature develop when biological fields act on existing organic units. The biofield is the basic unit of organic form and organization; molecules and cells are but "units of composition." Life evolves, according to Goodwin, on the interface between organism and environment, in a sacred dance generated in the interaction between organisms and the field in which they are embedded.

Goodwin did not affirm that biological fields would exist inde-pendently of living organisms. But others, such as Russian biologist V.M. Inyushin, did not hesitate to claim that biological fields are physically real, whether they are associated with an organism or not. According to Inyushin, such fields constitute a fifth state of matter, composed of ions, free electrons and free protons. Although in humans the field is attached to the brain, it may also project beyond an organism and produce telepathic phenomena.

English plant biologist Rupert Sheldrake, the author of a widely discussed if controversial biofield theory, has also been of the opinion that biofields have a reality of their own—existing apart from the organisms on which they act. In Sheldrake's view, "morphogenetic fields" are continuously shaped and reinforced by previously existing organisms of the same kind. Living members of a species are linked with the forms of past members of the same species through a causal link that transcends space and time. The linkage occurs by means of morphic resonance, a phenomenon requiring similarity of form or pattern. Resonance is reinforced by repetition, so that the more a given species has reproduced, the more it can reproduce in the future; the more a given behavioral routine has been learned by an animal, the faster other animals will learn that routine—and so on.

According to Sheldrake, morphogenetic fields do not carry a measurable form of energy. But the evidence regarding biofields speaks otherwise. Scientists at the A.S. Popov Bioinformation Institute in the former Soviet Union reported that the wave-length of the human bioenergy field moves in the range of 300–2,000 nanometers (where one nanometer is a billionth of a meter). They claimed that the field is linked with the effect produced by natural healers, whose field interacts with those of their patients. Investigators at Lanzhou University and at the Atomic Nuclear Institute in Shanghai, who had also researched the energy aspects of the human biofield, found that this field varies with the mental powers of the subject. Masters of Qigong, for example, have higher levels of bioenergies than most other people.

This observation has been borne out by the research of Valerie Hunt at Energy Fields Laboratory at UCLA. Using sophisticated equipment, both hard-wired to the test subjects and remote telemetry apparatus using short-range FM data transmission, Hunt measured the "emotional body" of people by attaching silver or silver-chloride sensors to various areas of their bodies. Her measurements show that the vibration frequencies of the energy radiated by the body encompass all the colors sensitives "see" in the human aura. Hunt found that the energy field radiating from mystics, seers and healers moves in a much higher frequency domain (around or beyond 400 Hertz) than the fields of persons in normal states of mind and body (the latter are usually below the 250 Hz range). Individuals of great spiritual gifts often register "aural" frequencies up to 200 kilohertz (*thousand* Hz)—the limit of Hunt's experimental apparatus. Those radiating at this level often report access to images and events that move far beyond the range of direct sensory experience, encompassing mystical archetypes, distant places, and remote times.

IN SUMMARY . . .

Universal interconnections correspond to a perennial intuition: mystics, poets as well as metaphysicians have always said that all things are connected—the petals of the flower in the garden with the stars of the farthest firmament. We now see that this is not merely of esthetic significance: it is also the precondition for creating a scientific theory with quasi-total vision. Such a theory builds its account of nature, life, and mind on the basis of subtle spatial and temporal interconnections, furnished by a holographically information-coding and -transmitting "fifth field."

A holographically-functioning super-weak field could exist in nature: physicists as well as biologists have found significant evidence of it. It now remains to "discover" it as a bone fide element of the universe. This revolutionary development marks the next milestone along the way of our continued explorations.

NOTES

1. The scientific definition of fields is more technical. A field is defined as a function of space and time, obeying an equation of partial derivatives with a definite variance. The result is a physical quantity that has different

values at different locations, with a mathematical function defining each position. The kinds of fields known to contemporary physics are vector, scalar, spinor, and tensor. They constitute electromagnetic fields, gravitational fields, fields of strong and weak nuclear force, and non-classical probability fields associated with the quantum state of elementary particles.

2. In the technical language of physics, Costa de Beauregard notes that the Lorentz and CPT invariance called for in any basic conceptualization (and formalization) of physical occurrences is required in the calculus of the joint probabilities of two physical occurrences as the expression of their interaction. This calculus must evidence the covariant propagation of information throughout space-time.

CHAPTER 13

DISCOVERING THE INTERCONNECTING FIELD

THE BASIC concept—the veritable kingpin—of genuine unified theories is universal interconnection. Indeed, the very possibility of such a theory hinges on finding the field in the universe that would connect atoms and galaxies, mice and men, brains and minds, and feed back information from each to all, and from all to each.

Is there reliable evidence that this "cosmic Internet" would truly exist in the universe? Obviously, it is not enough to just "postulate" the required field, tailor-made to the need for it. While this would be a simple procedure, it would not be science. Science must respect the law laid down by William of Occam in the 14th century: entities are not to be multiplied beyond necessity. Biologists are not free to advance a "life force" to explain why organisms carry out the functions associated with life, nor can psychologists put forward a "love force" to account for how people can love each other. In the same way, scientists cannot postulate a fifth field just to fill in gaps in the current texture of scientific knowledge. New entities (which can also be forces or fields) can only be postulated when doing so is the *simplest,* the most *economical,* and the most *rational* way of explaining a given set of findings or observations.

As it happens, to discover the interconnecting field the ad hoc postulation of a new entity is not necessary. Scientists can relate that field to one that is already known to exist in the universe.

INTRODUCING THE QUANTUM VACUUM

There is now a growing body of evidence that the interconnecting holofield is a specific manifestation of the cosmic quantum vacuum. But, just what *is* the quantum vacuum? The term seems mysterious, yet it refers to one of the most important, and as yet least understood, aspects of the physical universe. A deeper look at it is eminently worth our while.

In contemporary quantum physics the quantum vacuum is defined as the lowest energy state of a system of which the equations obey wave mechanics as well as special relativity. It is considerably more than just the state of a system, however: it is also the place where the mysterious "zero-point" field manifests itself. The energies of this field appear when all other, more conventional forms of energy vanish—at the zero-point (hence the name). Zero-point energies are "virtual" energies: they are not the same as the classical electromagnetic, gravitational, or nuclear forces. Rather, they are the very source of the electromagnetic, gravitational, and nuclear forces of the cosmos. As such, they are the originating source also of the energies that are bound in mass: the particles of matter that populate the known universe.

The technical definitions of the zero-point energy field that underlies the quantum vacuum indicate an almost infinite energy sea in which particles of matter are emergent substructures. According to English physicist Paul Dirac's calculations, all particles in positive energy states have negative-energy counterparts (by now such "antiparticles" have been found experimentally for all presently known particles). The zero-point energies of the quantum vacuum make up the "Dirac-sea": a sea of particles in the negative energy state. Though these particles are not observable, they are by no means fictional. By stimulating the negative energy states of the zero-point field of the vacuum with sufficient energy (of the order of 10^{27} erg/cm^3), a particular region of that field can be "kicked" into the real (that is, observable) state of positive energy. This process is known as pair-creation: out of the field come a positive energy (real) particle together with its antiparticle. Thus wherever there is matter, there is the Dirac-sea: the observable universe floats, as it were, on its surface.

Though the great majority of today's scientists are still relatively uninformed abut this mysterious, if entirely fundamental, energy domain, interest in it is growing rapidly. Important discoveries are coming to light. It is now known that it was the quantum vacuum that gave birth to the observable universe when a region of it (the Minkowski vacuum) became explosively unstable and split into matter and gravitation; and that it was this vast energy field that, in the subsequent and more sedately expanding Robertson-Walker phase of the universe, had synthesized the matter particles that subsist in space and time. It is further known that the quantum vacuum is not only the source, but also the sink of matter in the universe. Stephen Hawking's celebrated black-hole theory shows that at the "event horizon" of black holes one particle of the pair of particles synthesized in the vacuum escapes into surrounding space, while its antiparticle twin is sucked into the black hole, where it decays—back into the zero-point field of the vacuum.

This field contains a staggering density of energy. John Wheeler estimated its matter equivalent at 10^{94} g/cm^3. This, at first sight not strikingly impressive, quantity is in fact more than all the matter in the known universe. Compared with this level of energy, the energy density of the atomic nucleus—the most energetic chunk of matter in the cosmos—seems almost minuscule: it is "merely" 10^{13} g/cm^3.

If the energies of the vacuum's zero-point field were ordinary positive energies, the universe would instantly collapse to a size smaller than the head of a pin (or indeed the radius of an atom). This follows from Einstein's celebrated formula $E = mc^2$, which defines the equivalence of matter and energy. Real energy associated with mass corresponds to a specific amount of gravitation; thus the staggering vacuum energies, if real, would condense all the expanding stars and galaxies into a cataclysmic and entirely unimaginable "crunch."

But, at least until the final act of our universe (or of our present cycle of the universe), the world of matter—or, as we should now say, mass-energy—is safe from this ultimate catastrophe. Today, and for untold billions of years to come, the known universe will continue to float on top of this stupendous energy field. More exactly, it will continue to co-exist with it, rather like a set of bubbles suspended in it. Because, in terms of energy, the material world we know, and of which we are a part, is not a *solidification* of the vacuum's zero-point field, but actually a *thinning* of it.

CURRENT SPECULATIONS AND EMERGING INSIGHTS

A thin line divides what is already known and accepted about the quantum vacuum and what is still speculative and controversial. Since current knowledge is full of conceptual black holes, we shall no doubt see further insights emerging before long, as physicists track the mysteries and come up with new theories and devise new ways to test them. Here we review the most promising explorations, anticipating that, if not the final word, they could at any rate be on the right tack.

Consider the nature of the vacuum's zero-point field. Standard wisdom holds that it is homogeneous and isotropic—that is, the same in all directions in a uniform mix. This tenet derives from QED (quantum electrodynamics); it is favored by most physicists since it allows calculations of elegant and self-consistent mathematics. There are alternative approaches to this phenomenon, however, among them stochastic electrodynamics. Here the vacuum is seen as a field of constant quantum fluctuations; consequently the mathematics are more "messy." But what if the vacuum is indeed filled with a fluctuating energy field? If so, the alternative theory, though less simple and elegant, would be closer to the truth. This should give physicists pause for thought. Einstein said it well: we should make our theories as simple as possible—but not *simpler*.

That the field that underlies the motion and behavior of matter would have a structure of its own is not new to physics. Einstein's relativity theory also postulates a structured field: the space-time continuum. This field interacts with the real world of matter but, at least in the original interpretation, it does not have a reality of its own—it is purely geometrical. Lately a number of physicists have begun to question this assumption. Ignazio Licata in Italy and Manfred Requard in Germany, among others, developed theories of a relativistic universe in which space-time is not an abstract geometry but a physically real (so-called "reticular") field rooted in the quantum vacuum. Andrei Sakharov had suggested as early as 1968 that gravitation may be due to dynamic processes in the vacuum in the presence of matter. Hungarian physicist Lajos Jánossy followed up this suggestion by assigning the famous relativistic effects (such as the slowing down of clocks when they are accelerated close to the speed of light, or the increasing of the mass of objects at those velocities) to the interaction of real-world objects with the quantum vacuum. When accelerated close to the speed of light, the matter-particles of objects rub against the force-particles (bosons) of the vacuum, and this friction slows down their processes and increases their mass. In this alternative view the vacuum is not an abstract geometric structure, like Einstein's space-time, but a real physical field that interacts with the matter particles of the known universe.

Another Hungarian, maverick theoretician László Gazdag developed Jánossy's concept into a full-fledged "post-relativity theory." His theory, while still speculative, explains an otherwise puzzling fact. Why is it that, despite the enormous densities of the zero-point field energies, they are not ordinarily perceivable—or even measurable. Are these "virtual" energies entirely unobservable?

Gazdag says that this is not the case, as do Jánossy and other investigators. The vacuum field's energies are observable and indeed measurable, though not under all conditions. The vacuum's energy field (which, for the sake of simplicity we can henceforth describe as the energy of the vacuum) behaves as a kind of superfluid. Now, superfluids have curious properties. In super-cooled helium, for example, all resistance and friction ceases; it moves through narrow cracks and capillaries without loss of momentum. Conversely, objects move through the fluid without encountering resistance. (Since also electrons move through it without resistance, superfluids are also superconductors.) Thus, in a sense, a superconducting superfluid is not "there" for the objects or electrons that move through it—they get no information about its presence. Would electrons have measuring instruments, they would totally fail to register any trace of it.

Imagine, then, that the quantum vacuum is a superfluid with respect to the particles that move through it. The particles, and the objects built of them, do not register its presence; for them the vacuum does not exist. Since our bodies and brains are built of real-world positive-energy particles, and the ensemble of these particles moves through the vacuum as through a superfluid, our sense organs, and even our most sensitive instruments, do not register our passage. We may be pardoned for believing that there is no such thing as an energy sea surrounding us and our world.[1]

But the vacuum does not behave constantly and continuously as a frictionless superfluid. As noted by former Soviet physicist Piotr Kapitza (who devoted many years to the investigation of the properties of superfluid helium) in such a medium only those objects that are in constant quasi-uniform motion move without friction. If an object is strongly accelerated, vortices are created in the medium, these vortices produce resistance, and the classical interaction effects surface. For example, in the vortices of superfluid helium, strongly accelerated bits of wood or paper get carried along, much as in a classical fluid.

If a similar effect were to occur in the quantum vacuum, real-world particles that are not in constant and quasi-uniform motion would be affected by their movement through its energy field. This would yield the famous relativistic effects. It would also yield the more usual properties of real-world particles: inertia, gravitation, and electromagnetism.

In Gazdag's reinterpretation of Einstein's relativity theory the celebrated formulas describe the flow of bosons in the superfluid vacuum.[2] This flow is what determines the geometrical structure of space-time—and hence the trajectory of real-world photons and electrons. When light and matter particles move uniformly, space-time is Euclidean; and when they are accelerated (or else decelerated), the vacuum interacts with their motion. Then space-time appears curved.

Front-line research in physics confirms the basic notion that underlies these revolutionary assumptions. Current work follows up a suggestion made by physicists Paul Davies and William Unruh in the mid-1970s. Davies and Unruh, like Jánossy and Gazdag, based their argument on the difference between constant-speed and accelerated motion in the vacuum. Constant-speed motion would exhibit the vacuum's spectrum as isotropic (the same in all directions), whereas accelerated motion would produce a thermal radiation that breaks open the directional symmetry. The "Davies-Unruh effect," too small to be measured with physical instruments, prompted scientists to investigate whether accelerated motion through the vacuum would produce incremental effects. This expectation has borne fruit. It turned out that the inertial force itself could be due to interactions in the vacuum.[3]

In 1994 Bernhard Haisch, Alfonso Rueda and Harold Puthoff gave a mathematical demonstration that inertia can be considered a vacuum-based Lorenz-force. The force originates at the subparticle level and produces opposition to the acceleration of material objects. The accelerated motion of objects through the vacuum produces a magnetic field, and the particles that constitute the objects are deflected by this field. The larger the object the more particles it contains, hence the stronger the deflection—and the greater the inertia. Inertia is thus a form of electromagnetic resistance arising in accelerated frames from the distortion of the virtual-particle (and otherwise quasi-superfluid) gas of the vacuum.

Mass even more than inertia, also appears to be a product of vacuum interaction. If Haisch and collaborators are right, the concept of mass is neither fundamental nor even necessary in physics. When the massless electric charges of the vacuum (the bosons that make up the superfluid vacuum field) interact with the electromagnetic field, beyond the already noted threshold of energy, mass is effectively "created." Thus, astonishingly enough, mass may be a structure condensed from vacuum energy, rather than a fundamental given in the universe.[4]

If mass is a vacuum energy product, so is gravity. As we know from high school days, gravity is always associated with mass, obeying the "inverse square law" (it drops off proportionately to the square of the distance between the gravitating masses). Hence if mass is produced in interaction with the vacuum, then also the force that is associated with mass must be so produced. Indeed, the theory developed by Haisch and collaborators makes precisely this point. It gives a speculative, yet remarkably coherent mathematical account of the way gravity is produced in the zero-point field. The theory, which confirms Sakharov's pioneering insight, is based on the assumption that the electric component of this field causes charged particles to oscillate, and the oscillation gives rise to secondary electromagnetic fields. As a result, a given particle experiences both the electric forces of the zero-point field, which cause it to oscillate, and the secondary forces that are triggered in the field by another particle. The secondary field generated by the second particle acts back on the first particle. The net effect is an attractive force between the two particles. Thus gravitation turns out to be a long-range interaction among particles, much as the more familiar van der Waals force.[5]

Given the equivalence of the inertial and the gravitational forces, the theories of inertia and gravitation as the interaction of charged

particles with the vacuum stand or fall together. While these theories are not yet free of speculative elements, their implications have
much to recommend them. For one thing, gravitation no longer
appears as a mysterious force acting between two bodies distant
and separate from one another in space and time. In classical physics this force constitutes a metaphysical "action-at-a-distance,"
while in general relativity it is mediated by the geometry of space-
time. Action-at-a-distance is not an acceptable notion, and it is by
no means clear how a geometrical structure can create or convey
a physically real field (unless it is viewed as an ether-like space-
time medium—a view to which Einstein himself inclined in his later
years). In any case, gravitation cannot be given in the space-time
field itself, for if its staggering energy-density were associated with
gravitation (through the relativistic-equivalence of energy and
mass) the universe would instantly collapse to a size smaller than
the head of a pin. In the alternative ZPF/charged particle interaction theory this cannot occur, for the vacuum does not act upon
itself. Gravitation is not given in the zero-point field in the absence
of matter; it is only created there by the motion of particles. Hence
its value is limited to the mass of these particles, rather than extending to the mass-equivalent of all the bosons that make up the
vacuum's zero-point field.

In light of these fascinating leading-edge explorations, *all* the
fundamental characteristics we normally associate with matter are
vacuum interaction products: inertia, mass, as well as gravity.

Interactions between matter and the vacuum are discovered in
ever more domains. It is noted, for example, that under certain
conditions vacuum energies interact with the electrons that orbit
the atomic nuclei. These effects occur when electrons "jump" from
one energy state to another: the photons they emit exhibit the so-
called Lamb-shift (a frequency slightly shifted from its normal
value). Vacuum energies also create a radiation pressure on two
closely spaced metal plates. Between the plates some wavelengths
of the vacuum field are excluded, thereby reducing its energy

density with respect to the field outside. This creates a pressure—known as the Casimir effect—that pushes the plates inward and together.

Clearly, the quantum vacuum is not empty space; it is a significant element of the universe. Just how significant we can only guess, but even a conservative guess would admit that it goes far beyond the significance attributed to the vacuum in the classical theories. These theories have already accepted that the behavior of elementary particles is influenced by the vacuum, but they do not trace interactions between the macro-world of matter-energies to the quantum vacuum. Current research at the growing edge of science suggests, however, that interactions between the vacuum and the observable world of macro-level objects are embracing, and fundamental for our understanding of the nature of reality.

The work of a group of Russian physicists and biophysicists is particularly relevant here. As this writer learned on the occasion of his lectures at the Russian Academy of Sciences (in January 1996), Anatoly Akimov, G.I. Shipov, V.N. Binghi and co-workers developed a sophisticated theory of the "physical vacuum." In their theory the vacuum is a real physical substance extending throughout the universe: it registers and transmits the traces of particles and other objects. If validated by exhaustive testing (which has already begun), this theory could revolutionize physics for years to come.

Though abstract in conceptualization and mathematical in exposition, the claims of the Russian "torsion-field theory of the physical vacuum" are simple and basic. Essentially, the theory claims that all objects, from particles to galaxies, create vortices in the vacuum. The vortices created by particles and other material objects are information carriers, linking physical events quasi-instantaneously. The group-speed of these "torsion-waves" is of the order of 10^9 C—one billion times the speed of light. Since not only physical objects but also the neurons in our brain create and receive torsion-waves, it is not only particles that are "informed" of each other's presence (as in the famous EPR experiments), but also humans: our brain, too, is a vacuum-based "torsion-field trans-ceiver." This suggests a physical explanation of telepathy, remote viewing, and the other telesomatic effects discussed in Chapter Eight, together with the more standard (yet equally puzzling) phenomena of quantum nonlocality.

A vacuum-based signal-transceiving theory makes the universe far more interconnected than Einstein's relativity theory. The faster the signals travel from one point to another in space, the more points they connect in time. While there is still no absolutely simul-taneous interconnection between spatially separated events (that would call for signals traveling with infinite speed), when signals travel faster than light they open up the interaction cones that connect any given point in space with the past and the future of the universe. This can help explain why the structure of the cosmos is so uniform, even over the vast regions that could not have been connected by information traveling at the limited speed of light.

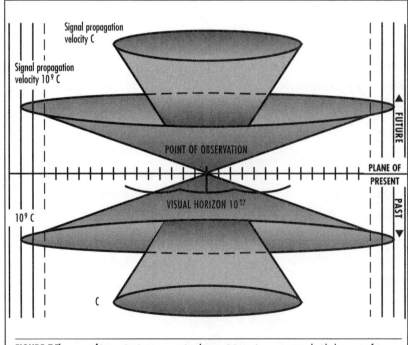

FIGURE 7 The range of interaction in a vacuum-signal transmitting universe, compared with the range of interactions in a light-velocity-limited universe. (scale illustrative only)

Of course, whether or not signals actually travel faster than light needs to be experimentally verified. The torsion-field vacuum theory of the Russians can allow this. The scientists have built a "torsion-field generator" operating at 60 gigahertz and in 1998 expect to send it into space. Installed aboard a Mars probe, the generator will send a torsion-wave from that planet toward Earth. If it were emitted at the same time as a light signal, it is predicted the torsion-wave would reach Earth fully eight minutes sooner.[6]

Torsion-waves are not only superluminal, they are also enduring. The metastable "torsion-phantoms" generated by spin-torsion interaction can persist even in the absence of the objects that generated them. The existence of these phantoms has been confirmed in the experiments of Vladimir Poponin and his team at the Institute of Biochemical Physics of the Russian Academy of Sciences. Poponin, who has since repeated the experiment at the Heartmath Institute in the U.S., placed a sample of a DNA molecule into a temperature controlled chamber and subjected it to a laser beam. He found that the electromagnetic field around the chamber exhibits a specific structure, more or less as expected. But he also found that this structure persists long after the DNA itself has been removed from the laser-irradiated chamber: the DNA's imprint in the field continues to be present when the DNA is no longer there. Poponin and his collaborators concluded that the experiment shows that a new field structure has been triggered from the physical vacuum. This field is extremely sensitive; it can be excited by a range of energies close to zero. The phantom effect is a manifestation, they claim, of a hitherto overlooked vacuum substructure.

Theories such as those we have cited in this chapter are truly revolutionary—they suggest a reconsideration of basic tenets in both relativity and quantum theory. The worldview implications of the theories are equally radical. In the emerging vision the very foundations of the universe acquire an active role in its functions and processes. Life, and even mind, is a manifestation of the constant if subtle interaction of the system of wave-packets classically known as "matter" with the underlying and interconnecting physically real vacuum field.

IN SUMMARY . . .

The interconnecting holofield is not likely to be a gravitational, electromagnetic, or nuclear field: it is more likely to be a "fifth field" in the universe. But, unlike classical fifth-field speculation, we need not suppose that it is a supernatural or esoteric phenomenon. Recent and as yet not widely known research at the leading edge of science indicates that it is a field created by the interaction of the quantum vacuum, the universe's fathomless energy sea, with the things and events of the observed and observable world.

The discovery of this field and its inclusion in the repertory of physically real events will make for a fundamental shift in the world picture projected by science. We shall trace the contours of this shift in the next and last chapter, as we contemplate the cosmic dance of matter, life, and mind in the whispering pond: in our subtly interconnected universe.

NOTES

1. Lack of measurement is frequently the reason why scientists refuse to believe in the existence of a phenomenon. Indeed, at the turn of the last century it was the lack of a measurable interaction between real-world objects and a space-filling substance through which they were said to move that compelled physicists to discard the idea of a "luminiferous ether."

2. The particles in question must be bosons rather than fermions; that is, their state (more exactly, wave-function) must not change on interaction. If it did, we would not have a uniform flow, but a structured system much as in the "real" world of space and time.

3. Inertia was originally defined as the property of a material object to either remain at rest or to move uniformly in the absence of external

forces—this is Newton's second law of motion $F = ma$: force equals mass times acceleration. Thus inertia appeared to be a fundamental quantitative property of matter. Yet it was a mysterious property: Newton himself could not fathom how it could be associated with material objects.

4. Though at first sight the new theory seems to contradict Einstein's famous mass-energy equivalence equation, it actually does not: energy is still equivalent to mass accelerated to the square of the speed of light. This equivalence means that mass is not only emerging *from* energy, but can also be reconverted *into* it: all one needs to do is to accelerate it to the speed of light, multiplied by itself. Though this speed cannot be attained by any mass in the universe, mass is still known to convert into pure energy—for example, in pair annihilation where a positron and an electron annihilate each other and give off the energy of their rest mass as gamma rays.

5. The interpretation offered by Haisch, Puthoff *et al* consists of two parts. In the first part the energy of the ultrarelativistic oscillations named *Zitterbewegung* by Schrödinger is equated to gravitational mass m_g, after dividing by c^2. Except for a factor of 2, this produces a relationship between the gravitational mass and electrodynamical parameters identical to the above postulated inertial mass m_i. However, Haisch, Puthoff *et al* show that the gravitational mass m_g, should be reduced by a factor of 2, thereby achieving a strict equivalence between m_i and m_g, i.e., between the forces of gravitation and inertia.

The second part of the analysis derives an inverse square force of attraction from the van der Waals force–like interaction between two driven oscillating dipoles. This analysis is admittedly incomplete, requiring further theoretical development in the framework of a fully relativistic model.

6. The theory, which at the time of writing has not been published outside Russia, is important and fascinating enough to merit a few technical details.

As we have already noted, the quantum vacuum is generally considered in the framework of quantum electrodynamics. This framework gives rise to elegant and relatively simple mathematics. But such formulas, though highly sophisticated, can be misleading nevertheless: they may not provide the best possible account of physical reality. In any case, quantum electrodynamics, as other scientific theories, can always be reconsidered or extended. This may be necessary in accounting for the phenomena that now come to light in regard to the quantum vacuum.

The Russian physicists do not hesitate to undertake this step. They take their cue from earlier work by Einstein. In a seminal treatment, G.I. Shipov shows that in accordance with the Clifford–Einstein program of the geometrization of space-time, the vacuum can be described not only in terms of Riemannian (four-dimensional) curvature, but also in terms of Cartan torsion. In the 1920s studies carried out by Albert Einstein and E. Cartan laid the foundation of the theory that subsequently became known as the ECT (Einstein–Cartan Theory). The idea stemmed originally from Cartan, who at the beginning of the century speculated about fields generated by angular momentum density. This idea was later elaborated independently by a number of Russian physicists, including N. Myshkin and V. Belyaev. They claim to have discovered the natural manifestations of enduring torsion-fields.

Presently Anatoly Akimov and his team consider the quantum vacuum as a universal torsion-wave carrying medium. The torsion-field is said to fill all of space isotropically, including its matter component. It has a quantal structure that is unobservable in nondisturbed states. However, violations of vacuum symmetry and invariance create different, and in principle observable, states.

The torsion-field theory takes a modified form of the original electron-positron model of the "Dirac-sea": the vacuum is viewed as a system of rotating wave-packets of electrons and positrons (rather than a sea of electron-positron pairs). Where the wave-packets are mutually embedded, the vacuum is electrically neutral. If the spins of the embedded packets have the opposite sign, the system is compensated not only in charge, but also in classical spin and magnetic moment. Such a system is said to be a "phyton." Dense ensembles of phytons are said to approximate a simplified model of the physical vacuum field.

When the phytons are spin-compensated, their orientation within the ensemble is arbitrary. But when a charge q is the source of disturbance, the action produces a charge polarization of the vacuum, as prescribed by quantum electrodynamics. When a mass m is the source of disturbance, the phytons produce symmetrical oscillations along the axis given by the direction of the disturbance. The vacuum then enters a state characterized by the oscillation of the phytons along their longitudinal spin-polarization; this is interpreted as a gravitational field (G-field). The gravitational field is thus the result of vacuum decompensation arising at its point of polar-ization—which is an idea that was originally introduced by Andrei Sakharov. Given that the gravitational field is characterized by longitudinal waves, it cannot be screened, which is in accordance with observation and experiment.

Hence m-disturbance produces the G-field, much as q-disturbance produces the electromagnetic field. We can now go further.

Following a thesis advanced by Roger Penrose, one can represent the vacuum equations in the spinor form and thereby obtain a system of nonlinear spinor equations where two-component spinors represent the potentials of torsion-fields. These equations can describe charged as well as neutral quantum and classical particles. We can thus allow that the vacuum is disturbed not only by charge and mass, but also by classical spin. In that event the phytons oriented in the same direction as the spin of the disturbance keep their orientation. Those opposite to the spin of the source undergo inversion; then the local region of the vacuum transits into a state of transverse-spin polarization. This gives the "spin field" (S-field), viewed as a condensate of fermion pairs.

As a result the Russian scientists can view the vacuum as a physical medium that assumes various polarization states. Given charge polarization, the vacuum is manifested as the electromagnetic field. Given matter polarization, it is manifested as the gravitational field. And given spin polarization, the vacuum manifests as a spin field. In this revolutionary theory *all* fundamental fields known to physics correspond to specific vacuum polarization states.

CHAPTER 14

THE COSMIC DANCE

WHEN SCIENTISTS recognize that the whole of the cosmos is constantly if subtly interconnected, they are in a position to offer a more coherent and less puzzle-ridden account of all the things that emerge in space and time, from atoms and galaxies to bacteria, mice, and men. Contemporary science approaches this new threshold of meaningfulness: the creation of a genuine unified theory is becoming a real possibility. It will afford us a well-nigh total vision of the known, and directly and instrumentally observable world.

QTV, quasi-total vision, promises to be fascinating; we can already glimpse some of its features. Here we sketch the basic landmarks, taking in turn the image of cosmos, matter, life, and mind.

A NEW VIEW OF THE COSMOS

In a universe where an underlying energy sea connects the observed phenomena, many of the paradoxes that baffle a universe of purely geometrical space-time can be overcome. New light is thrown first of all on the great puzzle of cosmology: why the universe seems so amazingly predisposed to life.

As noted in Chapter Five, the mystery concerns the precise tuning of the universe's basic constants. This predisposition of the physics of the universe to the biophysics and biochemistry of life encourages unusual flights of fantasy in the scientific community— it does not seem to allow a rational explanation.

Certainly, for Big Bang cosmology, why the physical constants of the universe were set the way they were at the cosmic birthing must remain inscrutable: the womb out of which the universe emerged is beyond the reach of the standard scenario. But this is not so for the latest multicyclic cosmologies. If the universe was not born in the Big Bang but only reborn in a "bang," we could know something of the womb from which it emerged. It may be that the cosmic womb was already "informed" by universes that preceded our own. Our universe could have inherited some features from previous universes.

Such an inheritance is entirely possible, provided that there was a physical medium that was able to hand down the characteristics of the "parent-universe" to our "offspring-universe." If current explorations are on the right tack, there *is* such a medium: it is the vacuum-based holofield. If that field was not created when our universe was born, and if it is a permanent memory of all the universes that were ever created, it could transmit the traces of prior universes to our own. Our universe would not have been born with an empty slate: the vacuum energy sea from which it had sprung would have been coded with the traces of previous universes.

Let us picture how this cosmic memory would operate. The co-evolution of the vacuum holofield with prior universes leads to the mutual harmonization of the processes of life with the physical preconditions that make life possible. The sacred dance of matter, life, and field acquires a cosmic dimension: in the course of inconceivable eons of time, the partners in the dance learn to take ever fancier steps, in ever closer harmony with each other.

In each successive universe atoms and molecules, cells and organisms become adapted to the basic constants that set the parameters of their evolution. The constants, in turn, become adjusted to the atoms and organisms that evolve in each universe. Thus at the explosive birth of each universe the memory-filled quantum vacuum produces precisely those small-scale deviations from the explosion's large-scale uniformities that create galaxies with stars, and stars with planets. And it synthesizes precisely that quantity of matter with precisely those forces of interaction that can bring forth molecules and cells and, on suitable planets, living organisms within entire biospheres.

This is an illuminating hypothesis. It tells us why our universe, born in a "bang" some 15 (or perhaps only 7 or 8) billion years ago, is so precisely predisposed to the evolution of life. It is because our universe came along as part of a perhaps long series of prior universes, unfolding within the enduring womb of an ongoing "metauniverse."

If this is so, can we inquire as to when the metauniverse itself was born—how long before our own universe?

A convincing answer to this question is beyond the ken of empirical science: it awaits the intuition of theologians and mystics. Yet science is not entirely at a loss in this regard: even if it cannot tell us just *when* the "primordial Big Bang" would have taken place, it can tell us something about *how* it would have taken place.

Here the theories of Russian cosmologist Andrei Linde become pertinent. The primordial Big Bang, he noted, could have been a reticular one: that is, it could have had several distinct regions. In this it would have resembled a soap bubble in which several smaller bubbles are stuck together. As such a bubble is blown up, the smaller bubbles become separated, each of them forming an independent bubble. This is what may have occurred at the primordial BB. That cosmic explosion would have had many regions, and each of them would have inflated into a distinct universe. The size and potential of the regions would have varied—most of them would have given birth to universes in which galaxies, stars, planets, and living beings are not possible. But among a large number of such "still-born" universes some may well have been viable. The universe in which we find ourselves would have been one such universe. Evidently "our" bubble was sufficiently large and "grainy" to have produced galaxies and stars, and some stars with planets and at least one planet with life. This could not be mere coincidence: we could never have evolved in any of the other universes.

The computations that flow from this theory of primordial inflation show that there is a high degree of probability that subsequent bubbles within a well-endowed region of the initial explosion are similarly endowed with evolutionary potentials. This would shift the puzzle of the lucky coincidence of characteristics to the primordial inflation.

We can go can do better than that, however: we can show that the present characteristics of the universe are not due to simple serendipity: they have *evolved* in the course of successive universes. We get this result if we add to the picture the vacuum-based holofield. As we have seen, this field is needed to explain how it is that distant parts of the universe show the same structures and the same patterns of development. Information traveling at the speed of light could not interconnect regions more than 15 billion light years from each other, but the waves propagating in the vacuum could: they travel much faster than light, perhaps—if the Russian torsion-field theories are right—as much as one billion times C, the light velocity. Thus the vacuum-based holofield "informs" all parts of the cosmos, ensuring consistency throughout its reaches.

We can specify that the holofield carries information from one "universe-bubble" to another: the periodic "bangs" do not destroy the vacuum's pre-existing substructure. This transcyclic memory in the underlying mega-universe ensures not only consistency throughout one and the same universe-cycle, but ongoing evolution *across* the cycles. As each baby universe receives the accumulated information of the prior parent universes, each emerging universe becomes more adapted to life than the one before it. The sequence of bang-created universes produces a learning curve. Our own universe, having come along within a string of prior universes, "in–formed" by the traces of its predecessors at its birth. No wonder it is so highly tuned to the requirements of life!

This vision of a self-creating and self-recording cosmos, evolving in a continuous if highly differentiated sweep from primeval unity to its present deeply linked diversity, corresponds to an intuition that was present in human consciousness from time immemorial. The creation myths of the widest variety of cultures agree that the things and beings of the observable universe came about as a concretization or distillation of the basic energy of the cosmos, descending from its original source. The physical world is a reflection of energy vibrations from more subtle worlds which, in turn, are reflections of still more subtle energy fields. This *leitmotif* runs through most mystical teachings. When we look horizontally outward from our physical body, science-writer and mystic John Davidson notes, we get descriptions at the physical level. But when we look vertically or inward, we find that the physical universe is a reflection downward of energy vibrations from more subtle worlds, which in turn are reflections of still more subtle worlds or energy fields. Creation, existence, is a progression downward and outward from the primordial Source.

In the Indian *Upanishads,* the original Source is an energy-dense space that came into being with the cosmos. This is *Akasha.* According to the account given by Swami Vivekananda, *Akasha* fills all space and gives rise to all that exists in it; it underlies and pervades air, fire, water, and earth. In the beginning there was only *Akasha* and at the end there will be only *Akasha. Akasha* becomes the Sun, the Earth, the moon, the stars and the comets; it also becomes the animal and the human body, the plants, and everything that exists. At the end of one phase everything will melt back into *Akasha,* to re-emerge from it in the next.

Prana, in turn, is the infinite and omnipresent power that acts on *Akasha. Prana* is motion, gravitation and magnetism; it is present in human action, in the nerve currents of the body, even in the force of thought. In the end all forces will resolve back into *Prana,* just as all things will lapse back into *Akasha.* And, most remarkably of all, the latter conserves the trace of all that has taken place in the universe. This is the "Akashic record": the enduring memory of the self-creating and self-recording universe.

The leading edge of physical cosmology also resurrects the ancient image of Vishna—like a lotus flower the manifest world unfolds, again and again, from the turbulent, liquid, creative substance, from the basic energy sea that creates and nourishes all things throughout space and time. This is much the same as saying that the universe emerges from the superfluid quantum vacuum and evolves throughout space and time, to relax back into it again—and then to emerge, again and again, in fiery pulses of cosmic creativity.

In its next development science will not need to call on divine intervention to explain why the cosmos is predisposed to life, nor will it have to rely on an almost inconceivable element of luck. The universe's hospitality to life is due neither to a special act of creation nor to blind chance: it is due to progressive cosmic evolution, unfolding across a long series of interconnected evolutionary cycles.[1]

A DIFFERENT VIEW OF MATTER

The Western commonsense view has always held that, in the final count, there are only two kinds of things that exist in the world: matter and space. Matter occupies space and moves about in it—it is the primary reality. Space is a backdrop or container. Unless it is furnished with material bodies, it can hardly enjoy a reality in itself.

This commonsense concept goes back to the Greek materialists; it was the mainstay also of Newton's physics. It has been radically revised in Einstein's relativistic universe (where space-time became an integrated four-dimensional manifold), and also in Bohr's and Heisenberg's quantum world. Now it may have to be rethought yet again.

The quasi-total vision of the new sciences suggests a further modification of this basic assumption about the nature of reality. We should no longer view matter as primary and space as secondary. It is to space—or rather to the field that occupies space—that we must grant primary reality.

Matter, as we have seen, is best viewed as a product of space—more exactly, of the vacuum's universal zero-point field that fills space. The seemingly solid objects that populate our world, and the flesh and bones that make up our body, are not constructed out of irreducible building blocks we could properly call "matter." The things we know as matter—and that scientists know as mass, with its associated properties of inertia and gravitation—are the results of subtle interactions in the depth of this space-pervading field. In the new vision there is no "absolute matter," only an absolute matter-generating energy field.

Physicists know that at the ultrasmall scale, "material" reality evaporates: particles, as isolated or isolatable entities, no longer exist; there are only quarks and the quantum fields they are embedded in. Quarks can only exist in a collective form within hadrons: protons, neutrons, and mesons. They cannot be separated from one another—it is not possible to produce a gas of quarks. Hence in the last count the atomic and molecular matter that makes up the "matter" part of our reality exists as various configurations of in themselves nonseparable quarks—and hence as fancy patterns superimposed on the underlying energy field. Consequently, material objects do not disport themselves within space as in a vessel: they are condensations, or critical nodes, of the space-filling vacuum energy field.

This new vision does not yet inform the thinking of the majority of scientists. If physicists were consistent they would view photons and electrons, and other quantum particles, as condensed quark-flows in superfluid space (more exactly, in the zero-point field of the quantum vacuum). But even particle physicists have difficulty in overcoming the standard view, which holds that photons are projected *across* space and *toward* technical equipment such as screens and mirrors—as, for example, in the famous double-slit experiment. The experimental apparatus is implicitly viewed as made up of sharply defined bodies, into which photons bump in various—and often very puzzling—ways. The primary reality remains the traveling particle and the material basis of the experimental apparatus. The space that lies between them, though known to be full of quantum fluctuations, is relegated to the status of secondary reality.

Yet scientists are no longer justified in thinking of photons and electrons as discrete entities projected across space and onto screens and mirrors. In a sound vision of physical reality even screens and other laboratory equipment are viewed as quantized waves in the underlying vacuum energy field. When scientists measure photons and electrons, they measure wave patterns in this field. When they conduct quantum experiments, one set of standing waves—the scientists themselves—experiments with another set of propagating waves—electrons and photons.

Although this kind of vision seems to stand common sense on its head, on closer scrutiny it turns out to be closer to everyday assumptions about the nature of reality than the standard conceptions of contemporary physics. For example, quantum fields are no longer purely ideal entities describing mere potentials—they are physically real entities interconnecting real-world particles and objects. Then the abstraction that boggles the mind of students in introductory physics classes is no longer there: light and gravitation are not phantom-like waves traveling in empty space. Space-time has not only a geometry, *à la* Einstein, but a basic physical reality. It is a plenum, a filled medium that can be perturbed—one that can create patterns and waves. Light and sound are travelling waves in this continuous energy field, and tables and trees, rocks and swallows, and other seemingly solid objects, are standing waves in it.[2]

It appears that the newest vision reconfirms the most ancient insight. The mystical notion that space is the creative source of matter is close to the truth. In the East, this notion goes back 5,000 years or more. According to the *rishis* (seers) of ancient India, space is not a mere framework for the adventures of the material things that alone are real. It is an eminent reality, a subtle substance that is just as real and perceptible as the four elements of air, fire, water, and earth. This view is reflected in the thinking of some contemporary Indian philosophers. Gopi Krishna, for example, the founder of the widely-known Kundalini Movement, said that the energies of the visible world originate from the primordial energy inherent in its creative potentials. The cosmos is like a boundless ocean dotted with icebergs. The cosmic ocean pervades space and time—it is the basis of all things. The ocean is impervious to our senses, but the gigantic ice-formations, transformed appearances of the underlying water, are perceptible. When we observe the world through our senses we see only the icebergs. But when we view reality internally, in *samadhi,* the icebergs vanish and water is perceived on all sides.

In the new-old concept of matter, Eddington's unflattering image of our spouse as a complicated set of differential equations can be replaced by a less abstract, though hardly a more flattering, image. The emerging QTV tells us that our spouse—as indeed ourselves, and all people and all living and nonliving things—is a complex standing wave in a cosmic sea of invisible, but physically real, energy.

ANOTHER VISION OF LIFE

The subtle relationship between the material things we meet with in our experience and the energy field that underlies them in the depths of the universe, transforms everything we know also of life and the world of the living. The emerging vision gives us the image of an interconnected web of nature that produces all the things that we observe in a continuous, organic process of self-creation.

The new vision tells us that the living organism is not the result of a series of accidents. The genetic information pool of the organism is not disjoined from its environment, and its variations are not the prey of mere chance. There is a direct though subtle link between the genome and the organism, and even the wider environment of the organism. Rather than a haphazard recombination of genes, the variations that bring forth new species are "adaptive": they are flexible responses on the part of the genetic substructure of the organism to the changes that it experiences in its milieu.

That such adaptive mutations would occur has been suggested time and time again by front-line thinkers and researchers. Although it conjures up the specter of Lamarckism (the doctrine, discredited decades ago, that the characteristics acquired by an organism in its lifetime can be inherited by its offspring), avant-garde investigators have often raised the possibility of a connection between the mutations of the genome and the demands of its milieu. The theory of adaptive mutations, which last provoked a whirlwind of debate and discussion in the 1980s, is currently under re-examination in the light of fresh evidence.

The emerging insight is not a throwback to previous, now outdated concepts—it does not maintain that the giraffe got its long neck because generations of giraffes stretched their necks to reach leaves on higher and higher branches. Instead, the new concept refers to the adaptability of the *genome,* the genetic information pool that codes long necks, the same as all other bodily features of the organism. The genome, microbiologists now find, is "fluid."

Controlled experiments have come up with numerous instances of environmental influences affecting the genome and provoking adaptive change in it. In flax, for example, directed change in the genome has been noted following treatment with fertilizers; various insects exposed to insecticides have been known to produce heritable amplifications of the specific genes that detoxify the chemicals and create resistance to the toxins; and similar modifications are known to occur in the genetic pool of a variety of species in consequence of electromagnetic and chemical influences.

It appears that the genome is "informed" of—or by—changing conditions in the environment. The isolation of the germline from the vicissitudes that befall the organism in its lifetime—one of the main pillars of classical Darwinism—is becoming undermined and will soon be abandoned. Organism and environment form part of a total system, and it is that system that evolves over time. Pure chance is bracketed out—even in regard to mutations in the genome, variations are said to occur within the context of a highly structured "epigenetic system."

It is now dawning on ever more biologists that germline stability is not due to the insulation of the genome, and that continued fitness is not the result of natural selection acting on mutations that are purely random. It is not contested that natural selection would play a role in evolution: variations that are disadvantageous for survival and reproduction do not persist, and this contributes to the observed fit between organism and environment. But natural selection is now viewed as a negative rather than a creative factor: it weeds out the nonfit mutants but does not ensure that mutants that are truly fit are created. The positive factor, biologists realize, is the close coupling of organism and environment within an embracing system that consistently self-evolves. This factor reduces the play of chance in evolution, linking the fluid genome with the systemic mutations that herald major evolutionary leaps.

The close coupling of genomes, phenomes, and environments emerges in ever more domains of biological interest. In place of the purely chance mutation–oriented view typical of classical Darwinism, and of the gene-oriented approach of the neo-Darwinian "modern synthesis," post-Darwinian developmentalists substitute a more sophisticated concept. Here evolution is more than the chance interplay of single factors: an entire array of factors and conditions conspires to adapt species to environments and environments to species, and to create new species to fill emerging environmental niches. The new concept focuses not on genes, individuals, or even groups as the primary units of selection, but on the functional relationships among a whole range of biological units from genomes to ecologies. The primary factor is not "fit" but "interaction": interaction between genes, between cells, between organisms and between all of them and the environment. Contact and communication are believed to issue in cooperation, and to culminate in synergy and symbiosis. This creates the total bio-ecological system that adapts and maintains itself, or else mutates and evolves. The focus shifts from "selfish genes" and organisms as mere vehicles for testing their self-centered purposes, to inherent drives within living systems at various levels of organization. The mechanism of evolution is essentially cooperative: it is "symbio-genesis" and "synergistic selection," rather than random mutation in isolated genomes.

The interconnections discovered by post-Darwinian developmentalists operate at all levels of life and evolution. They are often extremely subtle. For example, the transfers of energy among different parts of the organism are frequently so negligible that biochemists do not bother to measure them. What they measure instead is the "signal-release" (for example, from a particular organ through a hormone) to its binding at the receptor of another organ. A receptor at the cell-surface transduces the signal into an intracellular event, such as enzyme or gene activation. This is a clearly measurable event, produced by an energy transfer that, in itself, is hardly measurable. Similarly, subtle energy transfers produce measurable event not only within an organism, but also among organisms in entire populations and ecologies. This suggests the presence of continuous fields, linking all biological units within the great web of life of the biosphere.

The new vision of life, as a subtly linked web of units ranging from cells to ecologies, throws fresh light on the nature of our own bodies as well. We are not mere biochemical machines. This is a radical departure from the classical concept, still embraced by academic physiology and medicine, which sees the organism operating by reactions that relate its functions to its physiological structure, and its physiological structure to its body chemistry. In this outdated view, health depends on the integrity of the physiological structure, and this integrity depends in turn on balanced reactions involving a multitude of organic and inorganic chemicals. This means that whenever our body fails to function properly, the cause must be a structural defect arising from some chemical imbalance.

"Biochemical medicine" has been remarkably successful in a variety of applications, but this should not obscure the fact that it is not adequate in a variety of organic health situations. Another component needs to be added to treat the interactions that govern bodily function, structure, and chemistry. This component is a bioenergy field.

On a first view, the human bioenergy field (or biofield) appears to be electric and magnetic in nature. Neurophysiologists discovered that electric currents connecting certain regions in the brain produce the same effects as the injection of certain brain-stimulating chemicals. Other researchers found that properly administered electric currents stimulate cells to regenerate, enabling fractures to heal faster and tissues to self-repair. MRI scans have been added to X-rays and diathermy, testifying that electromagnetic fields play a nonnegligible role in the maintenance of the body's integrity. Imbalances in these fields indicate potential disturbances of body chemistry, prompting the breakdown of health. "Energy medicine" has joined biochemical medicine as a major complement.

But the electromagnetic biofield may not convey the full story: our bodies may be influenced by still more subtle energies. These can seldom if ever be detected directly, hence skeptical investigators have questioned their existence. Yet natural healers, as well as some physicians, make systematic use of them. Their experience demonstrates that subtle energies can influence the body's bioenergy field and thus, indirectly but sometimes crucially, its state of health.

Natural (or "alternative") medicine has made great strides in recent years. New developments include the establishment of the Office of Alternative Medicine at the National Institute of Health in Washington, the creation of a variety of professional journals, and a proliferation of books and conferences on research issues and clinical practice. Research is conducted on ways in which our consciousness interacts with our bodies, as in self-healing; ways in which one person's consciousness influences the consciousness and body of others through direct or indirect communication; and how such influence is transmitted "nonlocally" over space and time —the telesomatic effect noted in a growing number of studies and experiments.

Work in progress suggests that to the interactive chain that links function to structure to biochemistry in our bodies, and all of these to electromagnetic biofields, we must add yet another component. Traditionally called "etheric," "mental," or "spiritual," in the context of science's new vision the additional component is the holofield of the quantum vacuum. Human bodies, the same as other living organisms, are embedded in this field and are constantly interacting with it.

The ongoing dance of our body with the vacuum holofield changes our most fundamental notions of life and the living world. This is not the harsh domain of classical Darwinism, where each struggles against all, with every species, every organism and every gene competing for advantage against every other. Organisms are not skin-enclosed selfish entities, and competition is never unfettered. Life evolves, as does the universe itself, in a sacred dance with the underlying holofield. This makes living beings into elements in a vast network of intimate relations that embraces the entire biosphere—itself an interconnected element within the wider connections that reach out into the cosmos.

In the Earth's biosphere the network of subtle interrelations extends from the sequence of DNA in the chromosome of the cell to the global ecology as a whole. The genetic code in our bodies is not isolated from our life-sustaining environment; and one individual is not categorically separate from others. Subtle energies convey information on the dynamic structure of our physiology to every cell in our body, and from the dynamic processes that mold the environment to the genetic code within our cells. They also interlink our brain and body with the social and ecological systems in which we live.

In the emerging vision living organisms are linked with one another, interrelated by the holographically information-conserving and -transmitting field that pervades the universe. They all communicate together, dance the cosmic dance with each other.

THE OPEN-WINDOW CONCEPT OF MIND

In the cosmic dance life emerges out of non-life, and mind out of the higher domains of life. Once it has emerged, mind becomes an integral element of the dance—it is shaped by its connection with the rest of the universe, and in turn subtly shapes the rest.

This is an ancient concept, resuscitated in a new guise. For millennia, philosophers have wondered about the place of mind in nature. The theories have been many, but the alternatives they have contemplated are only a handful. To understand what is new and what is old about the current conception, we should sketch out the major alternatives.

Alternative 1. *Mind is a product of the brain—more exactly, a by-product of the survival functions the brain performs for the organism.* As organisms become more complex, they require a more complex "computer" to steer them so they can obtain the food, the mate, and the related resources they need in order to survive and reproduce. At a given point in this development mind appears. Thus mind is not the primary given in the real world; it is an "epi-phenomenon" that only appears as reality to those who possess sufficiently complex brains. This is the classic position of the *materialists*.

Alternative 2. *Mind is the ultimate reality; matter is but an illusion created by the human mind.* In the evolution of the universe mind was first, and it is still the first (and perhaps the only) reality. The material universe is but the creation of the human mind as it contemplates the—in its true nature, mental—world that surrounds us. This states the time-honored position of the *idealists*.

Alternative 3. *Mind and matter are both fundamental but entirely different; in humans they are associated through the brain.* The manifestations of mind cannot be explained by the systems that manifest it, not even by the staggeringly complex brains of human beings. In the case of humans, mind is associated with a material brain; but that brain is only the seat of mind and is not identical with it. When both matter and mind are recognized but are kept separate, we have the position of the *dualists*.

Alternative 4. *Matter and mind constitute a whole that cannot be taken apart, either in thought or in fact.* The entire distinction (introduced into Western thought by Descartes) between mind and matter is spurious: in the final analysis matter and mind form an integral whole. We must accept and treat them as such, regardless of where and in what form they are manifested. This is a relatively recent position: it is that of the *holists*.

Alternative 5. *Matter and mind are both real but they are not fundamental: they evolved together out of a still deeper level of reality.* The roots of both matter and mind extend into a deeper layer of reality that in itself is neither mental nor material.

The last-named position—Alternative 5—is the one that is at the heart of the vision that is now emerging in the sciences. It is a new view that does not yet have a well-defined name; we could best call it *evolutionism*. Certainly, it constitutes a dynamic conception that does not "reduce" reality either to inert, nonliving matter (as materialism), nor does it assimilate it to a mysterious, nonmaterial mind (as idealism). Both are held real, but (unlike in dualism), they are not viewed as basic elements of reality. Matter as well as mind have *evolved*—out of an entirely remarkable common womb: the zero-point energy field of the cosmic quantum vacuum.

The evolutionist view can be spelled out. When the stupendous process of self-creation got under way, matter and mind evolved together, to higher and higher, more and more complex forms. Even elementary particles had (and still have) some kind of proto-mind, and this mind gained in complexity and explicitness as the material systems that "carried" it—the atoms, molecules, cells, and organisms—became more complex and sophisticated. We humans experience as our personal consciousness the highly evolved mind that has coevolved with our sophisticated brain. Though on this planet this experience is uniquely explicit, it is not unique: all other organisms, and even molecules, atoms, and elementary particles, have some form of mental experience, with a level of explicitness that corresponds to their level of evolution.

There is a yet another element to be added to this conception. While that element is not unprecedented in the history of ideas, it is new in the sciences. It is the element of *interconnection*. The material/mental systems that evolve in the universe are constantly and intimately interlinked through the cosmic womb that has given birth to them. This womb—the quantum vacuum—is not a passive "has been" reality, but an active "nurturing" mother-factor that dances with all that it gives birth to.

The dance of our mind with the quantum vacuum links us with other minds around us, as well as with the biosphere of the planet and the cosmos that lies beyond it. It "opens" our mind to society, to nature, and to the universe. This openness has been known to mystics and sensitives, prophets and metaphysicians through the ages. But it has been denied by modern scientists and by those who took modern science to be the only way of comprehending reality. Now, however, the recognition of openness is returning to the new sciences. The emerging (and as yet revolutionary) insight is that the information conveyed by our brain regarding some features of the world beyond our cranium is not limited to the visible spectrum of electromagnetic waves and the audible spectrum of sonic waves. It extends to wave-propagations in the vacuum-based holofield. This field subtly interconnects our mind with the rest of the universe.

This insight has been forgotten in modern societies, very likely because everyday experience provides little evidence of it. This, however, is not because our wider connections are not real and functioning, but because evidence of them does not ordinarily penetrate to our normal waking consciousness. We exclude from our modern, commonsense awareness everything that does not conform to its modern, commonsense expectations.

This is not the case in traditional and non-Western societies. People in these societies often come up with remarkable forms of empathy, both with fellow humans and with nature. In the East the followers of Tao maintained that the highest good for man is to follow what is natural, and in the West native Americans voiced their oneness with all of nature. In the oft-cited words of Native American Chief Seattle, "This we know. All things are connected like the blood which unites one family. All things are connected. Whatever befalls the Earth befalls the sons of the Earth."

Such sentiments are in sharp contrast with the isolation felt by people in modern societies. Here our differences have been excessively accentuated over our bonds and commonalities. In the end our pursuit of individual self-fulfillment has led to the mistaken belief that we are categorically bounded by our skin, separate from the rest of society and from nature.

There have been exceptions, of course. Great poets such as John Donne and William Blake have sung of our oneness with the universe, and avant-garde scientists, such as Gregory Bateson and Arne Naess, have sought a detailed understanding of it. Einstein himself wrote, "a human being is part of the whole, called by us 'universe,' a part limited in time and space. He experiences his thoughts and feelings as something separate from the rest—a kind of optical delusion of his consciousness. This delusion is a kind of prison for us, restricting us to our personal decisions and to affection for a few persons nearest us."

The sense of separation that pervades modern societies does not plague people 24 hours a day. Though in ordinary waking consciousness most of us are mired in our separateness, trapped by the apparent discreteness of all things, when we sleep, meditate, or enter some other nonordinary state of consciousness, the situation changes. This is significant: the ordinary states of conscious awareness, though they seem all-encompassing, take up but a tiny part of our brain's activities.[3]

Nonordinary states of consciousness are not only real, but they are also accessible. William James noted in a famous statement penned over a hundred years ago, "Our normal waking consciousness . . . is but one special type of consciousness, whilst all about it, parted from it by the filmiest of screens, there lie potential forms of consciousness entirely different. We may go through life without suspecting their existence; but apply the requisite stimulus, and at a touch they are all there in all their completeness." People in "primitive" and classical cultures knew how to apply the requisite stimulus—some tribes, such as the !Kung Bushmen of the Kalahari desert, could enter altered states all at the same time. In many parts of the world ancient peoples combined chanting, breathing, drumming, rhythmic dancing, fasting, social and sensory isolation, even specific forms of physical pain to induce altered states. The native cultures of Africa and pre-Columbian America used them in shamanic procedures, healing ceremonies and rites of passage; the high cultures of Asia used them in various systems of yoga, Vipassana or Zen Buddhism, Tibetan Vajrayana, Taoism and Sufism; the ancient Egyptians used them in the temple initiations of Isis and Orisis; the classical Greeks used them in Bacchanalia, the rites of Attis and Adonis, and in the Eleusinian mysteries. Until the advent of Western industrial civilization, almost all cultures held such states in high esteem, for the remarkable experiences they could convey and their powers of personal healing and inter-personal contact and communication.

Today, at the leading edge of the contemporary sciences re-search on altered states of consciousness (ACSs) is becoming accepted as a legitimate part of the new discipline known as "consciousness research." Scientists know that such states can be induced not only by the classical shamanic and yogic practices and by psychedelic drugs, but even by simple breathing exercises (such as Grof's "holotropic breathing") and by the calm state induced by the suggestion of psychotherapists. Similar states do, of course, occur in deep prayer and concentration as well, and may also occur spontaneously—sometimes quite independently of the will of the person experiencing them.

The important fact to note about such states is that, as pioneer researcher Charles Tart noted, whatever their specific nature, they always tend to make our subtle connections to each other and to our environment more evident. This is true even in the state of dream-filled sleep. Already in mid-century, Carl Jung speculated that some of our dreams would reflect the collective unconscious of all humanity—they have a "numinous" quality about them. A similar view is now held by a number of psychologists. According to dream researcher Montague Ullman, even though we live as in-dividuals, separated from the larger whole of our species and our community, our dreams re-establish our connections; they further our efforts to live in harmony with nature and the universe. Unlike Freudian theories that speak of psychic entities in dreams at war with each other, the Ullman kind of dream theories relate dreaming to our interconnectedness with an embracing wholeness.

That the deep layers of our mind would connect us to one another is a view shared by physicist Fred Alan Wolf. He goes so far as to say that we are missing the point in looking for consciousness in individual brains. It may be, he wrote, that "my" consciousness does not exist only under my skin, but also out there, as an extended field.

The above conclusion, though daring, is strikingly confirmed in the work of psychiatrist Stanislav Grof. His "new cartography of the mind" (which, as we noted in Chapter Eight, includes a "transpersonal domain" in addition to the standard "biographic-recollective" domain) is based on countless experiences with altered states in patients. In session after session, when patients enter altered states they come up with information they could not have accessed through their eyes and ears. Grof concluded that in such states it is possible for us to get information from, and to identify with, practically anything in the universe. There are experiences of merging with another person into a state of dual unity, and also of fully assuming another person's identity. There are experiences of tuning into the consciousness of a group of people, as well as of expanding one's consciousness to such an extent that it encompasses all of humanity. One can entirely transcend the human limitations on experience and identify with what appears to be the consciousness of animals, plants, and even of inorganic objects and processes. It is also possible, according to Grof, to experience the consciousness of the entire biosphere, of the planet, and of the whole of the universe.[4]

Grof is not alone in making these claims, nor are the claims themselves new. They date back to the earliest roots of oriental philosophy; they were systematically described already in the *Yoga Sutras* of Patanjali. The ancient writings describe "the way" to harness one's mind to the forces of the universe: the art of yoga. Whoever follows the way, expands his or her consciousness without recourse to supernatural forces and entities—or even psychotherapists.

The mastery of the mind (*vibhuti*) described by Patanjali conveys astonishing powers and capacities. German consciousness researcher Franz-Theo Gottwald counted no less than 33 such items in the *Yoga Sutras*, ranging from a mastery over one's own senses to a mastery over the material world. The most frequently cited powers include a knowledge of the mind and thoughts of other beings; of the language of all living things; of the past and of the future; of hidden or distant objects; and of prior existences. The *siddhis* achieved this almost total mastery of body and mind, and the cosmically expanded consciousness that goes with it. Their *vibhuti* reached to a knowledge of the cosmos as a whole.

Some of the powers that were known to the *siddhis* are rediscovered by contemporary people in the practice of deep meditation. The experiences that surface in these states of mind and consciousness "clearly suggest," to quote Stanislav Grof, "that, in a yet unexplained way, each of us contains the information about the entire universe or all of existence, has potential experiential access to all its parts, and in a sense *is* the whole cosmic network . . ."

This claim is not without foundation. Though at present psychiatrists lack a scientific explanation for their findings, given the rapid pace of scientific discovery scientists may soon be in a position to give a meaningful account of some of these mystifying phenomena. The concept that emerges recalls what Jung had intuited: the human psyche is continuous throughout our species. Traffic between our interconnected brain/mind and the rest of the human world is constant, and it flows in both directions. We *send* our thoughts, impressions and emotions to others, and *receive* the thoughts, impressions and emotions of others. Everything that goes on in our mind leaves its wave-traces in the zero-point field of the quantum vacuum, and everything that goes on in our mind can be received by those who know how to "tune in" to the subtle traces that propagate there. As Vaçlav Havel, the intuitive writer elected to the presidency of the Czech Republic noted, it is as if something like an antenna were at our disposal, picking up signals from a transmitter that contains the experience of the entire human race.

We do have such an antenna in our body and, unlike in other species, in us that antenna is not a special receptor organ. Other species, too, pick up information from the fields that embrace the planet: fish navigate by means of the Earth's magnetic field—the intensity of this field depends on the direction in which they swim relative to the direction of the field—bees use the magnetic field in their orientation and communication; homing pigeons are influenced by flux densities down to a few nanoteslas in the magnetic field's fine fluctuations; and migratory birds fly at either right angles, or parallel, to the flux density lines of this field. But we humans are sensitive to the fields that surround us without requiring special organs. For example, scientists find that we react to electromagnetic signals and disturbances with a variety of symptoms that appear directly in our central nervous system (with displacements at 20-second intervals when subjected to atmospheric electromagnetic alternating fields between 10 and 50 kilohertz, and with perturbations of circadian rhythms, enzyme metabolisms and hormone production, among other things). Quasi-static magnetic and low frequency electromagnetic fields link up directly with processes in our organism, as transferred electrical information reaching the electro-mechanical (photon-phonon) code in the information transfer and storage mechanism of our nervous system.

Similarly, our brain could pick up information from the vacuum-based holofield without benefit of bodily receptors such as eyes and ears. Evidence indicates that space- and time-transcending information reaches our mind when we enter a free-ranging altered state of consciousness, such as the white-dream state between wakefulness and sleep, the state of deep meditation or prayer, and the special state produced by conscious breathing and systematic concentration.

Our constant, though not necessarily conscious, dance with other minds and the world aound us should give us a new sense of responsibility. Our thoughts and our feelings are not just our own concern: what we think and feel acts on others beyond the words we say and the attitudes we express. Our influence is subtle but nonetheless effective: as psychiatrists and psychotherapists well know, a person experiencing another person in the altered-state condition does not just recall that person and his or her experiences—he literally *becomes* that person, feeling his or her physical sensations, receiving his or her visual and other sensory perceptions and experiencing his or her emotions. Even in cases of less complete identification the impact tends to be striking, producing an impression that is indelibly recorded in the mind, subtly influencing thinking and feeling for the rest of one's life. Even when others do not experience our mental influence consciously, our thoughts and feelings can leave profound traces in their unconscious. We are, after all, connected with them through a constant two-way flow of images, thoughts, impressions and feelings, and these shape their minds whether they realize it or not.

AT THE FAR EDGE OF THE NEW VISION

Shortly after recovering from a nearly fatal illness, Gustav Fechner, the redoubtable founder of modern experimental psychology, penned the following statement. "When one of us dies, it is as if an eye of the world were closed, for all perceptive contributions from that particular quarter cease. But the memories and conceptual relations that have spun themselves round the perceptions of that person remain in the larger Earth-life as distinct as ever, and form new relations and grow and develop throughout all the future, in the same way in which our own distinct objects of thought, once stored in memory, form new relations and develop throughout our whole finite life." Could Fechner have hit upon an aspect of the truth?

Although here we enter territory that was traditionally the domain of metaphysics and mysticism, we can hazard an answer by extending the horizons of the new vision of science sketched in this

chapter. We can do so because at the leading edge of current scientific research a remarkable intuition may be receiving some form
of confirmation: our consciousness may be, in a sense, immortal.

The confirmation of this perennial intuition does not come directly from inspecting the contents of our own mind and consciousness, as in the mystical tradition, but from the possibility of
giving a scientifically valid explanation of the experiences that are
yielded by such introspection. The interconnecting field through
which our mind dances with the cosmos suggests an explanation—
it tells us that we would be wrong to dismiss the notion of immortality out of hand. Memories from apparent previous existences
may have a valid foundation after all—they could be information
accessed from a shared field of consciousness. Our feelings,
thoughts, and sensations are ongoingly read in, and preserved, in
the holographic spectrum of the quantum vacuum, and we gain
immortality already by leaving the traces of our body and mind in
that cosmic Akashic record.

There is, however, a further possibility. Could it be that the
experiences our mindful body reads into the cosmic information
pool are not distributed throughout that pool, but form an integrated set—much as a home-page on the World Wide Web? If
so, whatever we experience in our life, whatever thoughts, feelings,
or ideas flit through our consciousness, enter this location and
integrate there with all that has entered beforehand. Our own
"home-page" persists as a personal record throughout our lifetime
—and then beyond. Because, if information in the field does not
vanish when the things that produced them disappear but remain
conserved as "phantom patterns," the integrated record of the
experiences of our lifetime continues to exist *beyond* our lifetime.
And it could be accessible to whoever has the "code" to read it out
of this deep information pool.

Possibly, a fetus somewhere, growing in its mother's womb
would chance on (or perhaps be in some way predisposed for) the
code that could unlock the record of the experiences we have

accumulated in our own lifetime. It would begin to read out memories that were not its own, but ours. Its read-out would focus on the last additions to our lifetime record: the experiences that preceded (or accompanied) our death. Also the events that we ourselves had experienced with the greatest intensity would be "highlighted" in the record and become focal points in its read-out. These are the items that the fetus, and then the newborn and the growing child, would access and seem to recollect, in addition to its own experiences. Thus he or she would come into the world with memories of his or her own brief existence interlaced with memories of our own last hours or days, and of the traumatic or joyful experiences that had made a particularly deep impression on us.

These implications follow if information in the field is consistently integrated with prior information, much as on a Web site. If it *is,* we obtain a scientifically meaningful explanation of the oft-noted but hitherto mystical phenomena of karma and reincarnation.

We have now arrived at the deepest and most esoteric dimensions of human experience; at the outermost edges of the quasi-total vision emerging in the wake of the latest advances in the empirical sciences. That we could reach this far shore is significant in itself: it means that the separation between the natural sciences and the spiritual domains of experience is not permanent and irrecoverable. One day it may be bridged by the further advance of the scientific revolution that is unfolding before our eyes.

IN SUMMARY . . .

The paramount feature of the emerging quasi-total vision of cosmos, matter, life, and mind is subtle and constant interconnection. Evolution is not a blind groping toward nonexistent goals, a haphazard play with chance and accident. It is a systematic, indeed a systemic, development toward goals generated in the process itself. It unfolds because, as humankind has intuitively known throughout the ages, we, as all elements of the universe, are linked with one another. We are partners in a cosmic dance that is constant and unceasing. It "in–forms" our bodies and "in–forms" our mind. And when we allow it to penetrate our waking consciousness, it reinforces our sense of oneness with nature and the universe.

NOTES

1. This idea haunted the Russian physicist and Nobel laureate, A.D. Sakharov. It could be, he wrote, that after billions and billions of years of evolution something of the intelligence of the universe would survive the superdense conditions (of the Big Crunch) and inform the next universe. But, Sakharov admitted, he had never dared to express this thought in scientific publications. "Reflections," in *Science and Life*, No. 6, 1991, p. 29 (in Russian).

2. In the language of physics, photons and electrons are "spin-captured vectorial wave deformations" of the vacuum field, and screens and other solid bodies are "standing vectorial waves" in it. The former are propagating waves, like those that travel over the surface of the sea, and the latter are standing waves, like the ones created in a basin when water is flowing in and out at a constant rate. All material objects are standing waves; they are relatively stationary wave-patterns that only give the impression of being solid bodies.

3. A simple calculation shows the enormous difference between conscious processes and the overall capacity of the brain. The calculation is best carried out in reference to "bits," where one bit is the information contained in a yes/no answer to a question, or an either/or decision among two alternatives. It is customarily represented as the choice of a binary digit, 0 or 1. To encode or transmit one bit of information, the brain must have two potential states: 0 and 1. To encode or transmit two bits, it must have four potential states (00, 01, 10, and 11), and to encode or transmit three bits it must have eight potential states (000, 001, 010, 100, 110, 101, 011, and 111). The maximum information the brain can process in bits is equal to the logarithm at base 2 of the number of its possible states. Now, it is estimated that the processing of data from the senses requires about 10 billion bits per second. This calls for a truly astronomical

number of brain states, made possible by the network of 10 billion brain cells with a million billion connections. But information processing on the conscious level hardly ever involves more than about 10 bits per second. The rest of the processing occurs on the subconscious level—where the great majority of the messages that connect the brain with the world beyond are encoded and transmitted, as well as received and decoded.

4. The experience of "dual unity" is characterized by the loosening and melting of the boundaries of the body-ego and a sense of merging with another person in a state of oneness. In this experience, despite being fused with another, the experiencing person retains awareness of his or her own identity. In a related but distinct variety of experience the subject loses his or her own identity and has a sense of complete identification with the other. The "other" can be a living person, known from one's childhood or belonging to one's ancestry, or stemming from an apparent previous life. It can also be a famous personage from history, even a mythological or archetypal character. Identification with it involves body image, physical sensations, emotional reaction and attitudes, thought processes, memories, facial expression, typical gestures and mannerisms, postures, movements, even the inflection of the voice. The experience of group identification, in turn, involves a further extension of consciousness and melting of boundaries. Rather than identifying with individual persons, the experiencing subject has a sense of becoming an entire group of people with some shared racial, cultural, national, ideological, political, or professional characteristics. In the extreme case one can identify with the experience of all of humanity and with the human condition—its joy, anger, passion, sadness, glory, and tragedy.

NAMING THE FIELD:
A PROPOSAL FOR 21ST-CENTURY SCIENCE

W HAT SHALL we call the field that makes us, and all things in nature, organic parts in a subtly interlinked cosmos—in a cosmic whispering pond? If that field is a major, indeed a paramount, element in the universe, it deserves a name of its own. Describing it as the "vacuum-based zero-point holofield" is accurate but cumbersome; and the names with which other fields have been christened previously do not correspond to the insights now surfacing about the nature of this cosmic field.

What we are coming to recognize is that this field is both morphogenetic and morphophoretic; that is, it both *generates* and *carries* form. But it is more than a purely form-generating and -carrying entity: it is an interactive substructure in the most fundamental factor in the universe: in the quantum vacuum. That vacuum is "real" (even if speaking about a vacuum, which in ordinary parlance means empty space, as real appears to be a contradiction in terms), and it is omnipresent throughout space and time. Its holographic substructure "in–forms" the physical universe, the same as it "in–forms" the living world and the sphere of human mind and consciousness.

Since the interconnecting field is both a fundamental element in reality and a factor that enters into all our interactions with that reality, it deserves nothing less than a Greek symbol. We already have beta particles and gamma rays; alpha waves and omega factors. Why not call the vacuum-based cosmic holofield the "Ψ (psi) field"?

Why "Ψ" in particular? The most obvious answer would appear to be because it refers to—and perhaps explains—psi phenomena. This, however, is too easy: the universal holofield does considerably more than convey some varieties of extrasensory information: it also interconnects quanta and organisms, brains and minds and entire peoples and cultures. The rationale for using "Ψ" goes beyond parapsychology; it goes beyond psychology and neurophysiology, even biology and ecology. It embraces physics and cosmology, and the full range of the contemporary empirical sciences.

There is, indeed, a threefold rationale for identifying the vacuum-based holofield as the Ψ field:[1]

- *First:* In regard to the physical world, the field completes the description of the quantum state—it further specifies the wavefunction of the particle. The physical universe, complete with Ψ field satisfies Schrödinger's equation for the quantum state—$\Psi(x,t)$—much as the geometric structure of space-time satisfies Einstein's gravitational constant and the electromagnetic field satisfies Maxwell's equations.

- *Second:* With respect to the living world, the Ψ field is a factor of self-referentiality. It "in–forms" organisms consistently with their own and their milieu's morphology and may thus be viewed as a kind of intelligence—a generalized sort of "psyche" operating in the womb of nature.

- *Third:* In the domain of mind and consciousness, the Ψ field creates spontaneous communication between human brains as well as between human brains and the environment of the organisms that possess the brains. Though the field's effects are not limited to ESP and other esoterica, they convey the kind of information that has been traditionally subsumed under the heading of "psi phenomena."

A few years from now, given the vertiginous rate of progress and innovation in natural science in general and in the area of subtle interactions research in particular, we may well find that researching the Ψ field is just as acceptable and common as researching quarks and black holes today.

NOTE

1. The definition of the Ψ field was first suggested by the author in *The Creative Cosmos,* Floris Books, Edinburgh, 1993.

A CONCLUDING THOUGHT

T HE THEORIES and concepts of science are not merely the source of technological systems and gadgetry; they are also a source of meaning and, indirectly, a source of the values that we attach to meaning. When we come right down to it, how we relate to each other and to nature depends on our concepts of nature, of life, and of the thinking and feeling human being—on concepts that are involuntarily yet significantly influenced by science. If we believe that nature is a lifeless mechanism, a collection of passive rocks, we will come to believe that we are entitled to do with it as we please, so long as we do not go against our interests. Our choice of technologies will reflect these beliefs: we shall opt for powerful machines to extract, transform, use, and discard the energies and materials found in our environment. If we look on animals and other people as but more complex machines, we shall manipulate them, too: we will cut out their dysfunctional parts and organs, splice up their genes, or rewire the circuitry of their brain. We shall also manipulate people's social and political behavior, their labor, even their lifestyles, consumption patterns and leisure-time activities.

But what if nature—the universe itself—is not a passive rock or a lifeless machine? What if people are not complex machines, and not separate from each other and from their environment but profoundly, though subtly, linked? And what if the entire cosmos throbs with the creative energy of self-organization, constantly evolving, with periodic bursts of explosive innovation? If this is the concept we get from science, and if we assimilate it with our intellect and embrace it with our heart, would we still relate to each other and to our environment in quite the same way?

In this book we have argued that it is something like this organic image that science is now beginning to project. We have seen that the current wave of change sweeping through the natural sciences leaves behind the last remnants of the mechanistic view of life, mind, and universe. Space and time are united as the dynamic background of the observable universe; matter is vanishing as a fundamental feature of reality, retreating before energy; and continuous fields are replacing discrete particles as the basic elements of an energy-bathed cosmos. And the final destiny of this world need no longer be a lapse into the grayness of a lukewarm, empty and eternally unchanging nothingness, but could well be a cyclic self-renewal in a self-creating, self-energizing, and self-organizing mega-universe.

The current shift in science's concept of the world from a lifeless rock to an interconnected and quasi-living universe has intense meaning and significance for our times. The concept of a subtly interconnected world, of a whispering pond in and through which we are intimately linked to each other and to the universe, assimilated by our intellect and embraced by our heart, is part of humanity's response to the challenges that we now face in common. Our separation from each other and from nature is at the root of many of our problems; overcoming them calls for a recovery of our neglected, but never entirely forgotten, bonds and connections. Unexpectedly but perhaps not entirely accidentally, the vision emerging in the workshops of the avant-garde sciences could inspire ways of thinking and acting that would go a long way toward facilitating current efforts to transform the specter of a global crisis into the splendor of a humane and sustainable civilization.

With the poet's insight, T.S. Eliot asked, *"What are the roots that clutch, what branches grow out of this stony rubbish? Son of man, you cannot say, or guess, for you know only a heap of broken images . . ."* The new sciences help us surmount this predicament. They give us the vision of a whispering pond, of a universe where all things are linked in a fundamental unity. The insight that emerges is both meaningful and timely. It confirms psychologist-philosopher William James' image: we are like islands in the sea—separate on the surface, but connected in the deep.

Come,
sail with me on a quiet pond.
The shores are shrouded,
the surface smooth.
We are vessels on the pond
and we are one with the pond.

A fine wake spreads out behind us,
traveling throughout the misty waters.
Its subtle waves register our passage.

Your wake and mine coalesce,
they form a pattern that mirrors
your movement as well as mine.
Other vessels, who are also us,
sail the pond that is us as well;
their waves intersect with both of ours.
The pond's surface comes alive
with wave upon wave, ripple upon ripple.
They are the memory of our movement;
the traces of our being.

The waters whisper from you to me and from me to you,
and from both of us to all the others who sail the pond:

Our separateness is an illusion;
we are interconnected parts of the whole —
we are a pond with movement and memory.
Our reality is larger than you and me,
and all the vessels that sail the waters,
and all the waters on which they sail.

REFERENCES

CHAPTER ONE

For a general overview of the standard "Big Bang" theory and the major stages of cosmic evolution see, among others, Eric Chaisson, *Universe: An Evolutionary Approach to Astronomy,* Prentice-Hall, Englewood Cliffs, 1988; Eugene T. Mallove, *The Quickening Universe: Cosmic Evolution and Human Destiny,* St. Martin's Press, New York, 1987; and George Greenstein, *The Symbiotic Universe,* William Morrow, New York, 1987.

A highly readable account of black holes and the fate of matter is given in Stephen Hawking's *A Brief History of Time,* Bantam Books, New York, 1989.

The issue of the Omega point and the ultimate fate of the universe is discussed by John Gribbin in *The Omega Point: The Search for the Missing Mass and the Ultimate Fate of the Universe,* Bantam Books, New York, 1988.

CHAPTER TWO

The quotations regarding the enigmas of physical nature are from Arthur S. Eddington, *The Nature of the Physical World,* Macmillan, New York, 1929; from James Jeans, "Interview," in *Living Philosophers,* Simon & Schuster, New York, 1931, as well as "The Philosophy of Niels Bohr," in *Bulletin of the Atomic Scientist,* Vol. XIX, 7, 1971, and from Werner Heisenberg, *The Physicist's Conception of Nature,* Hutchinson, London, 1955, *Philosophic Problems of Nuclear Science,* Fawcett, New York, 1952, and "Development of concepts in the history of quantum theory" in *American Journal of Physics,* Vol. 43, 5, 1975.

Einstein's views in turn are reflected in *The Bohr-Einstein Letters,* Macmillan, London, 1971. The "smoky dragon" is presented in several of John Archibald Wheeler's writings, including "Bits, quanta, meaning," in *Problems of Theoretical Physics,* A. Giovannini, F. Mancini, and M. Marinaro (eds.), University of Salerno Press, Salerno, 1984.

CHAPTER THREE

The switch from classical physics and thermodynamics to the new nonequilibrium concept is stated in Ilya Prigogine, *Thermodynamics of Irreversible Processes,* 3rd. ed., Wiley-Interscience, New York, 1967, and discussed, in a more popular vein, in Ilya Prigogine and Isabelle Stengers, *Order out of Chaos: Man's New Dialogue with Nature,* Bantam Books, New York, 1984. Its implications for a new concept of evolution are spelled out by the present author in Ervin Laszlo, *Evolution: the General Theory,* Hampton Press, Cresskill, NJ, 1966; and by Sally J. Goerner, in *Chaos and the Evolving Ecological Universe,* The World Futures General Evolution Studies, Gordon and Breach, New York and London, 1994.

For Day's assessment of the acceleration of evolutionis in William Day, *Genesis on Planet Earth,* Yale University Press, New Haven, 1984.

CHAPTER FOUR

For the author's view on the brain/mind problem, see Ervin Laszlo, *Introduction to Systems Philosophy: Toward a New Paradigm of Contemporary Thought,* Gordon & Breach, New York, 1972, reprinted 1987; and *The Systems View of the World,* revised and enlarged edition, Hampton Press, 1996.

A highly readable account of the relation between perception and brain function in light of current neurophysiology is found *inter alia* in Jeremy Hayward, *Perceiving Ordinary Magic,* New Science Library, Shambhala, Boston and London, 1989.

Sir John Eccles offered the here cited view on memory in John Eccles and Daniel N. Robinson, *The Wonder of Being Human,* Shambhala, Boston and London, 1985, and Lashley's own conclusions regarding his experiments are in Karl Lashley, "The problem of cerebral organization in vision," *Biological Symposia,* Vol. VII, Visual Mechanisms, Jacques Cattell Press, Lancaster, 1942.

On Edelman's theory, see Gerald M. Edelman, *Neural Darwinism: The Theory of Neuronal Group Selection,* Basic Books, New York, 1987; and Gerald M. Edelman and V.B. Mountcastle, *The Mindful Brain: Cortical Organization and the Group-Selective Theory of Higher Brain Function,* MIT Press, Cambridge, MA, 1978. A more popular treatment is in Gerald M. Edelman, *Bright Air, Brilliant Fire,* Basic Books, New York, 1992. As regards memory in animals, see I.S. Beritashvili, *Vertebrate Memory: Characteristic and Origin,* Plenum Press, New York, 1971.

CHAPTER FIVE

The here quoted defense of the Big Bang theory is by P.J.E. Peebles, D.N. Schramm, E.L. Turner and P.G. Kron, "The case for the relativistic hot Big Bang cosmology," in *Nature,* August 1991, and on Jeans' views, see James Jeans, *Astronomy and Cosmogony,* Cambridge University Press, Cambridge, 1929.

The theory of quasi-steady state cosmology has been presented *inter alia* in F. Hoyle, G. Burbidge and J.V. Narlikar, "A quasi-steady state cosmology model with creation of matter," *The Astrophysical Journal,* Vol. 410, June 1993. The other multicyclic theory, known as the "self-consistent non-big bang cosmology," has been published in E. Gunzig, J. Geheniau and I. Prigogine, "Entropy and Cosmology," *Nature,* December 1987; and I. Prigogine, J. Geheniau, E. Gunzig, and P. Nardone, "Thermodynamics of cosmological matter creation," in *Proceedings of the National Academy of Sciences, USA,* Vol. 85, 1988.

The classic statement on the mystery of the constants and a possible explanation of it is in John D. Barrow and Frank J. Tipler, *The Anthropic Cosmological Principle,* Oxford University Press, New York, 1986, and Roger Penrose's calculations are given in his *The Emperor's New Mind,* Oxford University Press, New York, 1989, p. 340. Paul Davies' estimate is from his *God and the New Physics,* Simon & Schuster, New York, 1983, p. 168.

CHAPTER SIX

A concise and readable review of many of the puzzling quantum experiments is found in J.C. Polkinghorne, *The Quantum World,* Longman, London, 1984.

The EPR experiment, in turn, is stated in the classic study by Albert Einstein, Boris Podolsky, and Nathan Rosen, "Can quantum mechanical description of physical reality be considered complete?" in *Physical Review Letters,* Vol. 47, 1935; and Bell's equally classic analysis is in John S. Bell, "On the Einstein Podolsky Rosen paradox," in *Physics,* Vol. 1, 1964. Aspect's experiment is reported in a study by A. Aspect, P. Grangier, and G. Roger in *Physical Review Letters,* Vol. 49.9, 1982.

For a general discussion of the implications of Schrödinger's thought experiment see John Gribbin, *In Search of Schrödinger's Cat,* Bantam Books, New York, 1984, and Hegerfeldt's discovery of the problem with Fermi's calculations was reported in *Physical Review Letters,* Vol. 72, 1995.

Polkinghorne's remark on quantum theory is in J.C. Polkinghorne, *The Quantum World,* Longman, London 1984, p. 76.

The technical study of "Josephson effects" on living systems is in E. Del Giudice, S. Doglia, M. Milani, C.W. Smith and G. Vitiello, "Magnetic flux quantization and Josephson behaviour in living systems," in *Physica Scripta,* Vol. 40, 1989, and the principal "coincidences" in the build-up of matter in the universe are described by Sir Fred Hoyle in *The Intelligent Universe,* Michael Joseph, London, 1983.

CHAPTER SEVEN

For authoritative presentations of the theory of punctuated equilibria, see Niles Eldredge and Stephen J. Gould, "Punctuated equilibria: an alternative to phylogenetic gradualism," in *Models in Paleobiology,* Schopf (ed.), Freeman, Cooper, San Francisco, 1972; and Gould and Eldredge, "Punctuated equilibria: the tempo and mode of evolution reconsidered," in *Paleobiology,* Vol. 3, 1977. Further details are in Niles Eldredge, *Time Frames: The Rethinking of Darwinian Evolution and the Theory of Punctuated Equilibria,* Simon & Schuster, New York, 1985; Niles Eldredge, *Unfinished Synthesis. Biological Hierarchies and Modern Evolutionary Thought,* Oxford University Press, Oxford, 1985; and Stephen J. Gould, "Irrelevance, submission and partnership: the changing role of paleontology in Darwin's three centennials, and a modest proposal for macroevolution," in D. Bendall, (ed.), *Evolution from Molecules to Men,* Cambridge University Press, Cambridge 1983.

Denton's critique of Darwinist evolutionism is in Michael Denton, *Evolution: Theory in Crisis,* Burnett Books, London, 1986.

Jean Dorst and Etienne Wolff, as well as M. Schutzenberger and Giuseppe Sermonti, gave their comments in an interview conducted by Jean Staune in *Figaro Magazine,* 26 October 1991.

Walter Gehring's findings were reported in *Nature,* March 1995.

For Jacob's views, see François Jacob, *The Logic of Life: A History of Heredity,* Pantheon, New York, 1970.

CHAPTER EIGHT

General overviews of the new trends in consciousness research are given among others by M. Barinaga, "The mind revealed?," in *Science,* Vol. 249, 1990; and J. Gray, "Consciousness on the scientific agenda," in *Nature,* Vol. 358, 1992. More popular accounts are by John Horgan, "Can science explain consciousness" in *Scientific American,* July 1994; and David Freedman, "Quantum Consciousness," in *Discover,* June 1994.

The "classic" of the now considerable literature on near-death experiences (NDEs) is Elisabeth Kübler-Ross' *On Death and Dying,* Macmillan, New York, 1969. David Lorimer's comprehensive review of NDEs is *Whole In One: The Near-Death Experience and the Ethic of Interconnectedness,* Arkana, London, 1990. Regarding past-life experiences, the most authoritative work is that by Ian Stevenson, *Children Who Remember Previous Lives,* University of Virginia Press, Richmond, 1987. Famous if more controversial accounts are by Morris Netherton and Nancy Shiffrin, *Past Lives Therapy,* William Morrow, New York, 1978; Roger Woolger, *Other Lives, Other Selves,* Doubleday, New York, 1987; and Thorwald Dethlefsen, *Schicksal als Chance* (Fate As Destiny), Bertelsmann, Munich, 1979.

Elkin's anthropological findings are reported in A.P. Elkin, *The Australian Aborigines,* Angus & Robertson, Sydney, 1942.

The literature on transpersonal, telepathic, and related esoteric experiences is extensive; the studies most directly relevant here include Russell Targ and Harold Puthoff, "Information transmission under conditions of sensory shielding," in *Nature,* Vol. 251, 1974; Russell Targ and K. Harary, *The Mind Race,* Villard Books, New York, 1984; M. Ullman and S. Krippner, *Dream Studies and Telepathy: An Experimental Approach,* Parapsychology Foundation, New York, 1970; and M.A. Persinger and S. Krippner, "Dream ESP experiments and geomagnetic activity," in *The Journal of the American Society for Psychical Research,* Vol. 83, 1989.

Grinberg-Zylberbaum's experiments are reported in Jacobo Grinberg-Zylberbaum, M. Delaflor, M.E. Sanchez-Arellano, M.A. Guevara, and M. Perez, "Human communication and the electrophysiological activity of the brain," *Subtle Energies,* Vol. 3, 3, 1993.

Michael Murphy noted more than 25 studies that have shown that meditation can produce a synchronization of brain-wave activity between the right and left hemispheres as well as the anterior and posterior parts of the brain. See Michael Murphy, *The Future of the Body,* Jeremy Tarcher, Los Angeles, 1993.

For a general overview of telepathic experiments, see Harold E. Puthoff and Russell Targ, "A perceptual channel for information transfer over kilometer distances: historical perspective and recent research," *Proceedings of the IEEE,* Vol. 64, 1976; while the experiments with the "brain holotester" were carried out by Cyber, Ricerche Olistiche of Milan, and were reported in the journal *Cyber,* November 1992.

Daniel J. Benor has produced a comprehensive survey of distance healing experiments in *Healing Research,* Vol. 1, Helix Editions, London, 1993. Noteworthy contributions to this topic are Larry Dossey's books, *Recovering the Soul: A Scientific and Spiritual Search,* Bantam Books, New York, 1989; and *Healing Words: The Power of Prayer and the Practice of Medicine,* HarperSanFrancisco, San Francisco, 1993.

On telesomatic experiements, see among already numerous scientific studies, W. Braud and M. Schlitz, "Psychokinetic influence on electrodermal activity," *Journal of Parapsychology,* Vol. 47, 1983.

The basic study on the effect of meditation on a community is by Elaine and Arthur Aron, *The Maharishi Effect,* Stillpoint Publishing, Walpole, NH, 1986; while Randolph C. Byrd reported his pioneering experiments with prayer in *Southern Medical Journal,* Vol. 81.7, 1988.

Ignazio Masulli's research study, prepared at the behest of the author, is entitled "Analogies in some morphologies of ancient civilizations" *World Futures,* Vol. 48, 1–4, 1997.

Leonard Shlain's *Art and Physics* was published by William Morrow, New York, 1991. The experiment with "sympathetic magic" is reported in J.M. Rebman, Dean I. Radin, R.A. Hapke, and U.Z. Gaughen, "Remote influence of the automatic nervous system by a ritual healing technique" (mimeo), and Psychiatrist Stanislav Grof advanced his ideas on the new cartography of the mind in *The Adventure of Self-Discovery,* SUNY Press, Albany, 1988.

CHAPTER NINE

For a general overview of work in grand unification, see Barry Parker, *The Search for a Supertheory: From Atoms to Superstrings,* Plenum Press, New York, 1987. A more technical historical background has been furnished by Steven Weinberg in "The search for unity: notes for a history of quantum field theory," in *Daedalus,* Discoveries and Interpretations: Studies in Contemporary Scholarship (II), Fall, 1977.

Murray Gell-Mann published his quark theory in a technical paper, "A schematic model of baryons and mesons," in *Physics Letters,* Vol. 8.3, 1964, and restated it in a more accessible form in his Oppenheimer Memorial Lecture, *Elementary Particles,* The Institute for Advanced Studies, Princeton, October 1974.

CHAPTER TEN

Among Bohm's many writings, see the basic study *Wholeness and the Implicate Order,* Routledge & Kegan Paul, London, 1980. For a technical statement consult David Bohm and B.J. Hiley, "Non-relativistic particle systems," in *Physics Reports,* Vol. 828, 1986.

The citations from Werner Heisenberg are in *Daedelus,* 87, 1958, pp. 99–100, and his book, *Physics and Philosophy,* Harper & Row, New York, 1985, p. 54, while Henry Stapp's theory is stated in his book, *Matter, Mind, and Quantum Mechanics,* Springer Verlag, New York, 1993.

CHAPTER ELEVEN

Regarding Bohm's, Stapp's, and Prigogine's theories, see the notes to Chapter 10.

CHAPTER TWELVE

Sir Fred Hoyle's example is from *The Intelligent Universe,* Michael Joseph, London, 1983, and the "Twenty Questions" example was recalled by John Wheeler to the author in a personal communication.

Olivier Costa de Beauregard exposed his technical views at a lecture delivered at the Adriatico Research Conference on "Information theory in classical and quantum physics," 1995.

The citation from Shapley is from Harlow Shapley, "Life, hope and cosmic evolution," *Zygon,* 1 September 1966, and from Tiller in William A. Tiller, "Subtle energies in energy medicine," *Frontier Perspectives,* Vol. 4, 2, Spring 1995.

For Goodwin's concepts, see Brian Goodwin, "Development and evolution," *Journal of Theoretical Biology,* Vol. 97, 1982; "Organisms and minds as organic forms," *Leonardo,* Vol. 22,1, 1989. Inyushin's theories, in turn, are in the Russian-language textbook *Elementy teorii biologicheskogo polia,* published by the Kazakh State University, Alma Ata, 1978.

The full-length statement of Rupert Sheldrake's theories are in three books, *A New Science of Life,* Blond & Briggs, London, 1981; *The Presence of the Past,* Times Books, New York, 1988; and *The Rebirth of Nature,* Bantam Books, New York, 1991.

Hunt's experiments with the human biofield are reported in Valerie C. Hunt, *Infinite Mind,* Malibu Publications, Malibu, CA, 1996.

CHAPTER THIRTEEN

The complete theory by Gazdag is available to date solely in Hungarian: *A Relativitás Elméleten Túl* (Beyond Relativity Theory), Szenci Molnár Társaság, Budapest 1995. Aspects of it were published in English, however: László Gazdag, "Superfluid mediums, vacuum spaces," *Speculations in Science and Technology,* Vol. 12,1, 1989, and "Combining of the gravitational and electromagnetic fields," *ibid.,* Vol. 16,1, 1993.

The pathbreaking study on inertia is by Bernhard Haisch, Alfonso Rueda, and H.E. Puthoff, "Inertia as a zero-point–field Lorentz force," in *Physical Review A,* Vol. 49.2, February 1994, followed by lfonso Rueda and Bernhard Haisch, "Inertia as reaction of the vacuum to accelerated motion," in *Physics Letters A,* 240, 1998. For a more popular treatment, see "Beyond E = mc^2," by Haisch, Rueda and Puthoff, in *The Sciences,* November-December 1994. The theory of gravitation has been published by H.E. Puthoff, "Gravity as a zero-point fluctuation force," in *Physical Review A,* Vol. 39 (1989), and elaborated by Bernhard Haisch and Alfonso Rueda in "The zero-point flied and the NASA challenge to create the space drive," published in the *Journal of Scientific Exploration,* Vol. 11, 4, Winter 1997.

The full-length technical treatment of the vacuum torsion-field theory is G.I. Shipov's *A Theory of Physical Vacuum,* Moscow, 1995 (mimeo). An overview of the results of the numerous studies published in Russia is in Anatoly Akimov, "Heuristic discussion of the problem of finding long-range interactions." EGS-Concepts, Center of Intersectoral Science, Engineering and Venture, Non-Conventional Technologies (CISE VENT), Preprint No. 74, Moscow, 1991. Implications for the study of consciousness are in the collective work *Consciousness and Physical World,* CISE-VENT, Moscow, 1995 (in Russian, with English summaries). Reports on the "phantom DNA" are in P.P. Gariaev and V.P. Poponin, "Vacuum DNA phantom effect in vitro and its possible rational explanation," in *Nanobiology,* 1995; and in P.P. Gariaev, K.V. Grigor'ev, A.A. Vasil'ev, V.P. Poponin and V.A. Shcheglov, "Investigation of the fluctuation dynamics of DNA solutions by laser correlation spectroscopy," in *Bulletin of the Lebedev Physics Institute,* No. 11–12, 1989, pp. 23–30. Poponin and co-workers further suggest that a general explanation of the "phantom" phenomenon may be the nonlinear localized excitations (NLEs) found in an anharmonic Fermi-Pasta-Ulam lattice: see V.P. Poponin, "Modeling of NLE dynamics in one dimensional anharmonic FPU-lattice", in *Physics Letters A*, (in press).

CHAPTER FOURTEEN

For Swami Vivekananda's concept of *Akasha* and *Prana*, see his *Raja-Yoga*, published by Advaita Ashrama, Mayavati, Almora, University Press of India, 1937, while Gopi Krishna's here cited views are from "Kundalini for the new age," *The Odyssey of Science, Culture and Consciousness*, Kishore Gandhi (ed.), Abhinav Publications, New Delhi, 1990.

On adaptive mutations, consult *inter alia* the technical reports, R. Harris, S. Longerich, and S. Rosenberg, "Recombination in adaptive mutation," *Science*, Vol. 264, 8 April 1994; E. Culotta, "A boost for 'adaptive' mutation," *Science*, Vol. 265, 15 July 1994; and J. Radicella et al., "Adaptive mutation in Escherichia coli: a role for conjugation," *Science*, Vol. 268, 21 April 1995; as well as James Shapiro, "Adaptive mutation: who is really in the garden?", *Science*, Vol. 268, 21 April 1995. In turn, the post-Darwinian developmentalist concept of the role of natural selection is discussed in M.W. Ho and P.T. Saunders, (eds.), *Beyond Neo-Darwinism: Introduction to the New Evolutionary Paradigm*, Academic Press, London, 1984. For other aspects of the new biological paradigm, see *Evolutionary Processes and Metaphors*, M.W. Ho and S.W. Fox, (eds.), Wiley-Interscience, London, 1988; and Roberto Fondi, *La Revolution Organiciste*, Le Labyrinthe, Paris, 1986.

Einstein's insight about our mistaken self-perception is cited by Erwin Schrödinger in *What Is Life?*, Cambridge University Press, Cambridge, 1967.

William James' statement is from his *The Varieties of Religious Experience*, Modern Library, New York, 1902, 1929. For Ullman's view on the interpretation of dreams, see Montague Ullman, "Wholeness and dreaming," in *Quantum Implications: Essays in Honour of David Bohm*, B.J. Hiley and F. David Peat, (eds.), Routledge and Kegan Paul, London and New York, 1987. Fred Alan Wolfe's pronouncement on the extended nature of mind is in his "The Dreaming Universe", *Gnosis*, 22, Winter 1992.

The remarkable accounts of transpersonal experiences are from Stanislav Grof, *Beyond the Brain*, State University of New York Press, Albany, 1985. The repertory of astounding mental powers of the *siddhis* is given by Franz-Theo Gottwald, "Vibhuti oder Siddhi, Theory und Praxis der Erweiterung Menschlicher Fähigkeiten nach den *Yoga-Sutras des Patanjali*," in *Psyche und Geist*, Andreas Resch (ed.), Resch Verlag, Innsbruck, 1986, and Gustav Fechner's visionary statement is quoted by William James, in *The Pluralistic Universe*, Longmans, Green & Co., London, New York and Bombay, 1909.

FURTHER READING

A CONCISE LISTING, for the science-minded, of books and studies that pioneer or describe the current scientific revolution.

Aron, Elaine and Arthur, *The Maharishi Effect: A Revolution through Meditation*, Stillpoint Publishing, Walpole, NH, 1984

Barinaga, M., "The mind revealed?," *Science*, 249, 1990, 856–58

Barrow, John, *Theories of Everything*, Clarendon Press, Oxford, 1991

Barrow John D., and Frank J. Tipler, *The Anthropic Cosmological Principle*, Oxford University Press, London and New York, 1986

Bearden, Thomas E., *Toward a New Electromagnetics*, Tesla Book Co., Chula Vista, CA, 1983

Benor, Daniel J., *Healing Research: Holistic Energy Medicine and Spiritual Healing*, Helix Verlag, Munich, 1993

Bohm, David, *Wholeness and the Implicate Order*, Routledge & Kegan Paul, London, 1980

Bohm David, B.J. Hiley, and David Peat, *The Undivided Universe*, Routledge, London, 1993

Capra, Fitzjof, and David Steindl-Rast, *Belonging to the Universe*, HarperSanFrancisco, San Francisco,1983

Crick, Francis, *The Astonishing Hypothesis*, Charles Scribner, New York, 1994

Davidson, John, *Subtle Energy*, C.W. Daniel, Essex, 1983

—, *The Secret of the Creative Vacuum*, C.W. Daniel, Essex, 1989

Davies, P.C.W., and J.R. Brown (eds.), *The Ghost in the Atom*, Cambridge University Press, Cambridge, 1986

Davies, Paul, *God and the New Physics*, Simon & Schuster, New York, 1983

—, *The Mind of God*, Simon & Schuster, New York, 1992

—, and John Gribbin, *The Matter Myth*, Simon & Schuster, New York, 1992

Day, William, *Genesis on Planet Earth*, Yale University Press, New Haven, 1984

Del Giudice, E.G., S. Doglia, M. Milani, and G. Vitiello, in F. Guttmann and H. Keyzer (eds.), *Modern Bioelectrochemistry*, Plenum, New York, 1986

Delanoy, Deborah L., and Sunita Sah, "Cognitive and physiological psi responses to remote positive and neutral emotional states," in Dick Bierman (ed.), *Proceedings of Presented Papers*, American Parapsychological Association, 37th Annual Convention, University of Amsterdam, 1994

Dossey, Larry, *Recovering the Soul*, Bantam Books, New York, 1989

Duke, D.W., and W.S. Pritchard (eds.), *Measuring Chaos in the Human Brain*, World Scientific, London and Singapore, 1991

Eccles, John, and Daniel N. Robinson, *The Wonder of Being Human*, Shambhala, Boston and London, 1985

Edelman, Gerald M., *Bright Air, Brilliant Fire: On the Matter of Mind*, Basic Books, New York, 1992

—, "Morphology and Mind: is it possible to construct a perception machine?," *Frontier Perspectives*, 3, 2, 1993

Eldredge, Niles, *Unfinished Synthesis, Biological Hierarchies and Modern Evolutionary Thought*, Oxford University Press, Oxford, 1985

—, *Time Frames: The Rethinking of Darwinian Evolution and the Theory of Punctuated Equilibria*, Simon & Schuster, New York, 1985

—, and Stephen J. Gould, "Punctuated equilibria: an alternative to phylogenetic gradualism," Schopf, Freeman (eds.), *Models in Paleobiology*, Cooper, San Francisco 1972

Franz, Marie-Louise von, *Psyche and Matter*, Shambhala, Boston and London, 1992

Fröhlich, H. (ed.), *Biological Coherence and Response to External Stimuli*, Springer Verlag, Heidelberg, 1988

Gariaev, P.P., and V.P. Poponin, "Vacuum DNA phantom effect in vitro and its possible rational explanation," *Nanobiology*, 1995

Gazdag, László, *A Relativitás-elméleten Túl* (Beyond Relativity Theory), Szenci Molnár Társaság, Budapest, 1995

Gleick, James, *Chaos*, Viking, New York, 1987

Goodwin, Brian, "Development and evolution," *Journal of Theoretical Biology*, 97, 1982

—, "Organisms and minds as organic forms," *Leonardo*, 22, 1, 1989

Goswami, Amit, *The Self-Aware Universe*, Putnam, New York, 1993

Gould, Stephen J., "Irrelevance, submission and partnership: the changing role of paleontology in Darwin's three centennials, and a modest proposal for macroevolution," D. Bendall, (ed.), *Evolution from Molecules to Men*, Cambridge University Press, Cambridge, 1983

Gray, J., "Consciousness on the scientific agenda," *Nature*, 358, 277, 1992

Grimes, W., and K.J. Aufderheide, *Cellular Aspects of Pattern Formation: The Problem of Assembly*, Karger, New York and Basel, 1991

Grof, Stanislav, *Beyond the Brain: Birth, Death and Transcendence in Psychotherapy*, State University of New York Press, Albany, 1985

—, *The Adventure of Self-Discovery: Dimensions of Consciousness and New Perspectives in Psychotherapy and Inner Exploration*, State University of New York Press, Albany, 1988

—, with Hal Zina Bennett, *The Holotropic Mind*, HarperSanFrancisco, San Francisco, 1993

Gunzig, E., J. Geheniau, and I. Prigogine, "Entropy and Cosmology," *Nature*, 330, 6149, December 1987

Haisch, Bernhard, Alfonso Rueda, and H.E. Puthoff, "Inertia as a zero-point-field Lorentz force," *Physical Review A*, 49.2, February 1994

—, "Beyond E = mc²," *The Sciences*, November/December 1994

Hall, B.G., "Evolution on a petri dish," *Evolutionary Biology*, 15, 1982

Hansen, G.M., M. Schlitz, and C. Tart, "Summary of remote viewing research," in Russell Targ and K. Harary (eds.), *The Mind Race, 1972–1982*, Villard, New York, 1984

Harris, Errol E., *Cosmos and Anthropos*, Humanities Press, New York, 1991

Heisenberg, Werner, *Physics and Philosophy*, Harper & Row, New York, 1985

Herbert, Nick, *Elemental Mind*, Dutton, New York, 1993

Hiley, B.J., and David Peat, *Quantum Implications: Essays in Honour of David Bohm*, Routledge and Kegan Paul, London, 1987

Ho, M.W., "On not holding nature still: evolution by process, not by consequence," M.W. Ho and S.W. Fox (eds.), in *Evolutionary Processes and Metaphors*, Wiley-Interscience, London, 1988

—, "The role of action in evolution," *Cultural Dynamic*, 4, 1991

Honorton, C., R. Berger, M. Varvoglis, M. Quant, P. Derr, E. Schechter, and D. Ferrari, "Psi-communication in the Ganzfeld: experiments with an automated testing system and a comparison with a meta-analysis of earlier studies," *Journal of Parapsychology,* 54, 1990

Hoyle, Fred, *The Intelligent Universe,* Michael Joseph, London, 1983

Hoyle, Fred, G. Burbidge and J.V. Narlikar, "A quasi-steady state cosmology model with creation of matter," *The Astrophysical Journal,* 410, 20 June 1993

Huxley, Aldous, *The Perennial Philosophy,* Harper & Row, New York, 1970

Ives, Herbert, "Extrapolation from the Michelson-Morley experiment," *Journal of the Optical Society of America,* 40, 1950

—, "Revisions of the Lorentz transformations," *Proceedings of the American Philosophical Society,* Vol. 95, 1951

Jacob, François, *The Logic of Life: A History of Heredity,* Pantheon, New York, 1970

Jahn R.G., and B.J. Dunne, "On the quantum mechanics of consciousness, with application to anomalous phenomena," *Foundations of Physics,* 16, 8, 1986

Jansen, B.H., and M.E. Brandt (eds.), *Nonlinear Dynamical Analysis of the EEG,* World Scientific, London and Singapore, 1993

Jung, Carl G., "Commentary on The Secret of the Golden Flower," in R. Wilhelm, *The Secret of the Golden Flower,* Harcourt, Brace & World, New York, 1962

Kafatos, M. (ed.), *Bell's Theorem, Quantum Theory and Conceptions of the Universe,* Kluwer, Dordrecht, 1989

—, and Robert Nadeau, *The Conscious Universe,* Springer Verlag, New York, 1990

Kauffman, Stuart, *The Origins of Order: Self-Organization and Selection in Evolution,* Oxford University Press, Oxford, 1993

Krippner, S., W. Braud, I.L. Child, J. Palmer, K.R. Rao, M. Schlitz, R.A. White, and J. Utts, "Demonstration research and meta-analysis in parapsychology," *Journal of Parapsychology,* 57, 1993

Laszlo, Ervin, *The Interconnected Universe,* World Scientific, London, New Jersey and Singapore, 1995;

—, *The Creative Cosmos,* Floris Books, Edinburgh, 1994

—, *Introduction to Systems Philosophy,* Gordon and Breach, New York, 1972

—, Evolution, The General Theory, Hampton Press, Cresskill, NJ, 1996

Licata, Ignazio, "Dinamica Reticolare dello Spazio-Tempo" (Reticular dynamics of space-time), *Inediti* 27, Soc. Ed. Andromeda, Bologna, 1989

Lindley, David, *The End of Physics,* Basic Books, New York, 1993

Lorimer, David, *Whole In One: The Near-Death Experience and the Ethic of Interconnectedness,* Arkana, London, 1990

McKenna, Terence, Rupert Sheldrake and Ralph Abraham, *Trialogues at the Edge of the West,* Bear & Co., Santa Fe, NM, 1992

Moody, Jr., Raymond A., *The Light Beyond,* Bantam Books, New York, 1988

—, *Life After Life,* Mockingbird Books, Covington, 1975

Netherton, Morris and Nancy Shiffrin, *Past Lives Therapy,* William Morrow, New York, 1978

Pagels, Heinz, *The Cosmic Code,* Bantam Books, New York, 1990

Peat, F. David, *Einstein's Moon,* Contemporary Books, Chicago, 1990

Penrose, Roger, *The Emperor's New Mind,* Oxford University Press, New York, 1989

Persinger, Michael A., and Stanley Krippner, "Dream ESP experiments and geomagnetic activity," *The Journal of the American Society for Psychical Research,* 83, 1989

Polkinghorne, John, *Reason and Reality,* Trinity Press, Philadelphia, 1991

—, *The Quantum World,* Longman, London, 1984

Pribram, Karl, *Brain and Perception: Holonomy and Structure in Figural Processing*, The MacEachran Lectures, Lawrence Erlbaum, Hillsdale, NJ, 1991

Prigogine, I., J. Geheniau, E. Gunzig, and P. Nardone, "Thermodynamics of Cosmological Matter Creation," *Proceedings of the National Academy of Sciences, USA*, 85, 1988

Prigogine, Ilya, *Thermodynamics of Irreversible Processes*, Wiley-Interscience, New York, 1967

—, "Why irreversibility? The formulation of classical and quantum mechanics for nonintegrable systems," *International Journal of Quantum Chemistry*, 1994

Prigogine, Ilya, and Isabelle Stengers, *Order Out of Chaos: Man's New Dialogue with Nature*, Bantam Books, New York, 1984

Puthoff, Harold A., "Source of vacuum electromagnetic zero-point energy," *Physical Review A*, 40.9, 1989

Rein, Glen, "Modulation of neurotransmitter function by quantum fields," *Planetary Association for Clean Energy*, 6,4, 1993

—, "Biological interactions with scalar energy-cellular mechanisms of action," *Proceedings of the 7th International Association of Psychotronics Research*, Georgia, December 1988

Requardt, Manfred, "From 'matter-energy' to 'irreducible information processing'— Arguments for a paradigm shift in fundamental physics," K. Haefner (ed.), *Evolution of Information Processing Systems*, Springer Verlag, New York and Berlin, 1992

Rosen, Joe, *The Capricious Cosmos*, Macmillan, New York, 1991

Saunders, Peter T., "Evolution without natural selection," *Journal of Theoretical Biology*, 1993

Schroeder, Gerald L., *Genesis and the Big Bang*, Bantam Books, New York, 1990

Sheldrake, Rupert, *A New Science of Life*, Blond & Briggs, London, 1981

—, *The Presence of the Past*, Times Books, New York, 1988

Stapp, Henry P., *Matter, Mind, and Quantum Mechanics*, Springer Verlag, New York, 1993

Stevenson, Ian, *Unlearned Language: New Studies in Xenoglossy*, University of Virginia Press, Charlottesville, 1984

—, *Children Who Remember Previous Lives*, University Press of Virginia, Charlottesville, 1987

Stewart, Ian, *Does God Play Dice?*, Blackwell, Cambridge, 1992

Talbot, Michael, *The Holographic Universe*, HarperCollins, New York, 1991

Targ, Russell, "Remote viewing replication: evaluated by concept analysis," Dick J. Bierman (ed.), *Proceedings of Presented Papers*, Parapsychological Association 37th Annual Convention, University of Amsterdam, 1994

—, "What I see when I close my eyes," *Journal of Scientific Exploration*, 8,1, 1994

—, and Harold Puthoff, "Information transmission under conditions of sensory shielding," *Nature*, 252, 5476, 1974

Tart, C.T., "Physiological correlates of psi cognition," *International Journal of Parapsychology*, 5, 1963

Tiller, William A., "Subtle energies in energy medicine," *Frontier Perspectives*, 4 2, Spring 1995

Trefil, James, *Reading the Mind of God*, Anchor Books, New York, 1989

Ullman, Montague, "Wholeness and dreaming," Hiley and Peat (eds.), in *Quantum Implications*, op. cit.

Varvoglis, Mario, "Goal-directed and observer-dependent PK: an evaluation of the conformance-behavior model and the observation theories," *The Journal of the American Society for Psychical Research*, 80, 1986

Weinberg, Steven, *The First Three Minutes*, Basic Books, New York, 1988

—, *Dreams of a Final Theory*, Pantheon Books, New York, 1992

Wheeler, John Archibald, "Quantum cosmology," L.Z. Fang and R. Ruffini (eds.), *World Science*, Singapore, 1987

—, "Bits, quanta, meaning," A. Giovannini, F. Mancini, and M. Marinaro (eds.), *Problems of Theoretical Physics*, University of Salerno Press, Salerno, 1984

White, John, *The Meeting of Science and Spirit*, Paragon House, New York, 1990

Wigner, Eugene, I.J. Good (ed.), *The Scientist Speculates*, Heinemann, London, 1961

Wilber, Ken, (ed.), *Quantum Questions*, Shambhala, Boston and London, 1985

Woolger, Roger, *Other Lives, Other Selves*, Doubleday, New York, 1987

Zohar, Danah, *The Quantum Self*, William Morrow, New York, 1990

Zukav, Gary, *The Dancing Wu Li Masters*, Bantam Books, New York, 1989

INDEX